Why Intelligent Design Fails

Why Intelligent Design Fails

A Scientific Critique of the New Creationsim

EDITED BY

MATT YOUNG

TANER EDIS

RUTGERS UNIVERSITY PRESS

New Brunswick, New Jersey, and London

Third printing, 2005

Library of Congress Cataloging-in-Publication Data

Why intelligent design fails : a scientific critique of the new creationism /
    edited by Matt Young and Taner Edis.
        p. cm.
Includes bibliographical references and index.
     ISBN 0-8135-3433-X (hardcover : alk. paper)
    1. Creationism.   2. Evolution.   I. Young, Matt, 1941- II. Edis, Taner,
1967-
   BL240.3.W49 2004
   213—DC22

                           2003020100

A British Cataloging-in-Publication record for this book is available from the British
Library

The publication program of Rutgers University Press is supported by the Board of
Governors of Rutgers, The State University of New Jersey.

Manufactured in the United States of America

It is interesting to contemplate an entangled bank, clothed with many plants of many kinds, with birds singing on the bushes, with various insects flitting about, and with worms crawling through the damp earth, and to reflect that these elaborately constructed forms, so different from each other, and dependent on each other in so complex a manner, have all been produced by laws acting around us.

Charles Darwin,
On the Origin of Species, 1859

Contents

Preface

This book concerns *intelligent design*, a new and comparatively sophisticated form of creationism. Concerned scientists and educators will find answers to questions such as the following:

> *What is intelligent-design (ID) creationism?* A conservative religious
>   agenda masquerading as a scientific alternative to evolution.
> *Why is it universally rejected by mainstream science?* Because it makes no
>   real predictions and lacks explanatory power.
> *What are the specific scientific errors in intelligent design arguments?* They
>   ignore how modern science has already solved the questions they
>   raise about complexity.

The book is critically needed today, when state school boards are under pressure to allow the teaching of intelligent-design creationism in the public schools, sometimes at the option of the local boards. Intelligent design is presented as *science*; therefore, effectively answering its claims depends on the availability of resources highlighting the *scientific* shortcomings of ID claims.

This book is addressed primarily to scientists, science educators, and educated people who are interested in the intelligent-design controversy and want to understand why mainstream scientists almost universally reject intelligent design. We present critiques of intelligent design from a scientific perspective yet at a level that is accessible to readers who do not have specific expertise in some or all of the disciplines of our contributors. Our readership will thus include science educators, readers of popular science books, students, and scientists who do not have direct expertise with intelligent-design arguments.

Critical Responses

Convinced that they are excluded from science classes by sheer prejudice, intelligent-design proponents have concentrated in particular on attacking what they perceive to be the prior naturalistic commitments of modern science. Since the intelligent-design movement has gained the sympathies of a number of philosophers with conservative religious backgrounds, including eminent figures such as Alvin Plantinga, book-length critical responses have largely concentrated on the philosophical and theological issues raised by intelligent design.

Such critical books generally presume that the primary mistake in intelligent-design claims is philosophical and argue that non-naturalistic ideas should not be given scientific consideration at all. A concentration on philosophy is understandable, given that intelligent design has been scorned by the mainstream scientific community. Nonetheless, intelligent design presents itself as a respectable *scientific* alternative to Darwinian evolution and natural selection. Hence, although they intend to exclude intelligent design altogether, philosophical critiques emphasizing naturalism in science have paradoxically given intelligent design a measure of intellectual legitimacy despite its overwhelming scientific failure. Too often, intelligent design has become a philosophical perspective to be debated in typically inconclusive fashion, with only passing reference to the decisive answers from mainstream science.

The integrity of science education is best supported by presenting the successes of actual science rather than highlighting philosophical attempts to define the boundaries of proper science. Intelligent design, like older versions of creationism, is not *practiced* as a science. Its advocates act more like a political pressure group than like researchers entering an academic debate. They seem more interested in affirming their prior religious commitments than in putting real hypotheses to the test. They treat successful scientific approaches—for example, a preference for naturalistic explanations—as mere prejudices to be discarded on a metaphysical whim.

Pointing out such dubious practices in the intelligent-design camp must remain an important part of any critique. It is even more important, however, to show how mainstream science explains complexity much more successfully, even without invoking a mysterious intelligent designer. We know how Darwinian mechanisms generate information. We know how evolutionary biology fits in with our modern knowledge of thermodynamics. We know that computer science and information theory give creationism no comfort. In the end, scientists reject claims of intelligent design because of their failures, not because intelligent design is indelibly stamped with a philosophical scarlet letter.

This book therefore emphasizes the scientific failures of intelligent design. It makes a strong case against intelligent design from many disciplines and also demonstrates its technical failures to readers who are not experts in the specific fields being discussed. This book will be a standard reference for anyone seriously interested in the debate over intelligent design because, in the end, science education, not philosophy, is the primary area in which the battle over the new creationism is being fought.

Contributors

The contributors to this book are legitimate academic researchers who are also active in criticizing varieties of creationism, particularly intelligent design. Some have more than 40 years of experience, all are published authors, and some have published one or more books in their fields. They have also published articles, both in print journals and on the Internet, that refute the neocreationists' pretensions. Indeed, we have recruited them because of our familiarity with their work criticizing intelligent design.

The contributors represent a broad spectrum of physicists, mathematicians, and computer scientists, as well as biologists. This spectrum is necessary because, although some intelligent-design advocates, such as Michael Behe, concentrate on biology, the ID movement is not ultimately about biology alone. Its claims also have direct relevance to physics and cosmology, and its principal arguments are mathematical. Indeed, much of the intelligent-design literature concerns physics and cosmology and, particularly in William Dembski's work, computer science and mathematics. The main purpose of the movement is to reestablish *design* as a basis for explaining our world. All our sciences are thus under attack.

Defending science from such attacks requires not just direct criticisms of intelligent-design arguments but also explanations of how contemporary science has made significant progress in explaining complexity. This defense calls for an interdisciplinary approach because complexity is an area of research that draws together different perspectives from physics, computer science, and biology. Our contributors therefore include representatives from these disciplines and others. They emphasize the success of mainstream approaches to the evolution of complex systems in order to demonstrate the scientific emptiness of intelligent design.

MATT YOUNG, BOULDER, COLORADO
TANER EDIS, KIRKSVILLE, MISSOURI

Acknowledgments

The editors thank, in particular, Wesley Elsberry for initiating and maintaining the list server whose members formed the backbone of this book. In addition to the authors of individual chapters, we thank Erik Tellgren and Richard Wein, who offered pithy criticisms in the book's early stages. We are also grateful to Jason Rosenhouse and Jeffrey Shallit, who read and commented on the entire manuscript, as well as Barbara Forrest and Glenn Branch.

We are further indebted to our editors, Audra Wolfe and Adi Hovav, for sticking with us through an arduous review and to Amy Bix for originally getting us in touch with Audra Wolfe. Dawn Potter copyedited the manuscript with care and precision. Adrian Mellott and Kevin Padian offered excellent suggestions that led to several new chapters.

Taner Edis thanks Scott Ellis, head of the science division at Truman State University, for encouraging his studies of the borderlands of science. Matt Young is especially indebted to Béla Scheiber, president of the Rocky Mountain Skeptics, for originally getting him interested in intelligent-design creationism, and to Jim McNeil, head of the physics department at the Colorado School of Mines, for his encouragement.

Finally, the editors thank each other and all their co-authors for a year or so of hard work and good-natured criticisms.

Matt Young dedicates his share of this volume to his grandchildren, Alexandra and Noah, and advises them never to put the cart before the horse or the conclusion before the evidence.

Taner Edis dedicates his share to his parents, who are the source of his love for science.

Why Intelligent Design Fails

Introduction

MATT YOUNG AND TANER EDIS

INTELLIGENT DESIGN is the successor to old-fashioned creationism but dressed in a new coat—its hair cut, its beard trimmed, and its clock set back 10 or 15 billion years. It is nevertheless a hair's-breadth away from creationism in its insistence that everyone is wrong but its proponents, that science is too rigid to accept what is obvious, and that intelligent-design advocates are the victims of a massive conspiracy to withhold the recognition that their insights deserve.

Creationism, though very popular in its young-earth version, has failed as a strategy for introducing religious beliefs into the science curriculum. Enter neocreationism, or intelligent design. Not as obviously a religious conceit as creationism, intelligent-design creationism has made a case that, to the public, appears much stronger. Pertinently, its proponents are sometimes coy about the identity of their designer. They admit to the age of the earth or set aside the issue, and some even give qualified assent to pillars of evolutionary theory, such as descent with modification. They have therefore been able to feign a scientific legitimacy that creationism was never able to attain.

This aura of legitimacy has enabled the proponents of intelligent design to appeal to the public's sense of fairness and ask that intelligent design be added to school curricula, alongside Darwinian evolution, as an intellectually substantial alternative. Intelligent design, however, has found no support whatsoever from mainstream scientists, and its proponents have not established a publication record in recognized and peer-reviewed scientific journals. They have nevertheless raised a significant sum of money and embarked on a single-minded campaign to inject intelligent design into the science curriculum.

Intelligent-Design Neocreationism

Biblical literalism, in its North American form, took shape in the 1830s. One impetus was the attack on slavery by religious abolitionists. Slave owners or their ministers responded by citing biblical passages, notably Genesis 9:24–27, as justification for enslaving black people: "And Noah awoke from his wine, and knew what his younger son [Ham, the supposed ancestor of black people] had done to him. And he said, Cursed be Canaan [son of Ham], a servant of servants shall he be unto his brethren. . . . Canaan shall be [Shem's] servant . . . and [Japheth's] servant" (King James version).

At about the same time, the millennialist strain in Christianity began a resurgence in Britain and North America. This movement, the precursor of modern fundamentalism, also stressed the literal truth of the Bible (Sandeen 1970). Most millennarians and their descendants, however, adjusted their "literal" reading of Genesis to accommodate the antiquity of the earth. Some accepted the *gap theory*: that God created the heavens and the earth in the beginning but created humans after a gap of millions or billions of years. Others accepted the *day-age theory*, which recognized the days mentioned in Genesis as eons rather than literal 24-hour days. There was, therefore, no contradiction between science and their religious beliefs. Many evangelical thinkers went as far as to accept not only an old earth but even biological evolution, provided that evolution was understood as a progressive development guided by God and culminating in humanity (Livingstone 1987).

Evolution education did not become a fundamentalist target until the early twentieth century. Then, in the aftermath of the Scopes trial, literalist Christianity retreated into its own subculture. Even in conservative circles, the idea of a young earth all but disappeared (Numbers 1992).

The pivotal event behind the revival of young-earth creationism was the 1961 publication of *The Genesis Flood*, co-authored by hydraulic engineer Henry M. Morris and conservative theologian John Whitcomb. Morris resurrected an older theory called *flood geology* and tried to show that observed geological features could be explained to be results of Noah's flood. In Morris's view, fossils are stratified in the geological record not because they were laid down over billions of years but because of the chronological order in which plants and animals succumbed to the worldwide flood. To Morris and his followers, the chronology in Genesis is literally true: the universe was created 6000 to 10,000 years ago in six literal days of 24 hours each. With time, Morris's *young-earth creationism* supplanted the gap theory and the day-age theory, even though some denominations and apologists, such as former astronomer Hugh Ross, still endorse those interpretations (Numbers 1992, Witham 2002).

Creationists campaigned to force young-earth creationism into the biology classroom, but their belief in a young earth, in particular, was too obviously religious. A few states, such as Arkansas in 1981, passed "balanced-treatment" acts. Arkansas's act required that public schools teach *creation science*, the new name for flood geology, as a viable alternative to evolution. In 1982, Judge William Overton ruled that creation science was not science but religion and that teaching creation science was unconstitutional. Finally, the 1987 Supreme Court ruling, *Edwards v. Aguillard*, signaled the end of creation science as a force in the public schools (Larson 1989).

The intelligent-design movement sprang up after creation science failed. Beginning as a notion tossed around by some conservative Christian intellectuals in the 1980s, intelligent design first attracted public attention through the efforts of Phillip Johnson, the University of California law professor who wrote *Darwin on Trial* (1993). Johnson's case against evolution avoided blatant fundamentalism and concentrated its fire on the naturalistic approach of modern science, proposing a vague "intelligent design" as an alternative. Johnson was at least as concerned with the *consequences* of accepting evolution as with the *truth* of the theory.

In 1996, Johnson established the Center for Science and Culture at the Discovery Institute, a right-wing think tank. In 1999, the center had an operating budget of $750,000 and employed 45 fellows (Witham 2002, 222). Johnson named his next book *The Wedge of Truth* (2000) after the wedge strategy, which was spawned at the institute. According to a leaked document titled "The Wedge Strategy" (anonymous n.d.), whose validity has been established by Barbara Forrest (2001), the goal of the wedge is nothing less than the overthrow of materialism. The thin edge of the wedge was Johnson's book, *Darwin on Trial*.

The wedge strategy is a 5-year plan to publish 30 books and 100 technical and scientific papers as well as develop an opinion-making strategy and take legal action to inject intelligent-design theory into the public schools. Its religious overtone is explicit: "we also seek to build up a popular base of support among our natural constituency, namely, Christians. . . . We intend [our apologetics seminars] to encourage and equip believers with new scientific evidence's [sic] that support the faith" (anonymous n.d.).

Johnson remains a leader of the movement, although he is the public voice of intelligent design rather than an intellectual driving force. That role has passed to Michael Behe, William Dembski, and others.

Intelligent Design in Biology

The intelligent-design movement tries to appeal to a broad constituency, drawing on widely accepted intuitions about divine design in the world (see chapter 1). As the wedge document acknowledges, however, reaching beyond conservative Christian circles has been a problem. Success evidently requires a semblance of scientific legitimacy beyond lawyerly or philosophical arguments.

Thus, intelligent design has gathered steam with the publication of biochemist Michael Behe's book *Darwin's Black Box* (1996), which argues that certain biochemical structures are so complex that they could not have evolved by natural selection. Behe calls such complex structures *irreducibly complex*.

An irreducibly complex structure is any structure that includes three or more parts without which it cannot function. According to Behe, such a structure cannot have evolved by chance because it cannot function with only some of its parts and more than two parts are not likely to form a functioning whole spontaneously. Behe identifies, for example, the bacterial flagellum and the blood-clotting system as irreducibly complex. To prove his point, he relies heavily on the analogy of a mousetrap, which he says cannot function with any one of several parts missing. Behe's argument founders, however, on the pretense that the irreducibly complex components came together at once and in their present form; he makes no effort to show that they could not have coevolved. Chapter 2 shows that Behe's mousetrap is a failed analogy designed to hide this likelihood.

Many intelligent-design neocreationists accept what they call *microevolution* but reject *macroevolution*. That is, they accept the fact of change within a species but reject the idea that a species may evolve into a new species. Chapter 3 shows that their assignment of living organisms into kinds is incoherent and that there is no substantive difference, no quantitative demarcation, between microevolution and macroevolution. The distinction is wholly arbitrary and fragments the tree of life, whereas common descent provides a neat and compact picture that explains all the available evidence.

Chapter 4 shows that the scientific evidence Behe presents is equally flawed. Behe discounts the importance of the fossil record and natural selection and adopts a belief in a designer outside nature because of the concept of an irreducibly complex system, which he cannot defend. He further points to a supposed absence of scientific articles describing the evolution of biochemical systems deemed to be irreducibly complex and a paucity of entries for the word *evolution* in the indexes of biochemistry textbooks. Behe is a legitimate scientist, with a good record of publication. Nevertheless, his claims, which he likens to the discoveries of Newton and Copernicus, are not well regarded by most biologists, and they are reminiscent of standard God-of-the-gaps arguments.

Chapters 5 and 6 develop the theme introduced in chapter 4. Chapter 5 explains how an irreducibly complex structure can readily evolve by exapting existing parts and then adapting them to new functions. These new functions take form gradually, as when a feathered arm that originally developed for warmth turns out to be useful for scrambling uphill and only gradually adapts for flying. Chapter 5 details precisely how such exaptation-adaptation gradually formed the avian wing.

The eubacterial flagellum is one of the favorites of the intelligent-design proponents and occupies a place in their pantheon that is analogous to the place of the eye in the creationist pantheon. Chapter 6 shows that the flagellum is by no means an "outboard motor" but a multifunctional organelle that evolved by exaptation from organelles whose function was primarily secretion, not motility. It is not irreducibly complex.

Chapter 7 links the previous chapters to those that follow. It shows how the laws of thermodynamics do not preclude self-organization, provided that there is energy flow through the system. In addition to energy flow (an open system), self-organization requires only a collection of suitable components such as atoms or molecules, cells, organisms (for example, an insect in an insect society), and even the stellar components of galaxies, which self-organize through gravitational energy into giant rotating spirals. Using two examples, Bénard cells and wasps' nests, chapter 7 demonstrates how complex structures can develop without global planning.

Intelligent Design in Physics and Information Theory

Behe passed the torch to mathematician and philosopher William Dembski, who claims to have established a rigorous method for detecting the products of intelligent design and declares further that the Darwinian mechanism is incapable of genuine creativity. Hiding behind a smoke screen of complex terminology and abstruse mathematics, Dembski in essence promulgates a simple probabilistic argument, very similar to that used by the old creationists, to show that mere chance could never have assembled complex structures. Having failed to convince the scientific community that his work has any substance, Dembski nevertheless compares himself to the founders of thermodynamics and information theory and thinks he has discovered a fourth law of thermodynamics (Dembski 2002, 166–73).

Dembski has gone well beyond Behe with a mathematical theory of specified complexity. According to Dembski, we can establish whether or not an object or a creature was designed by referring to three concepts: contingency, complexity, and specification.

> *Contingency.* Dembski looks to contingency to ensure that the object could not have been created by simple deterministic processes. He would not infer intelligent design from a crystal lattice, for example, because its orderly structure forms as a direct result of the physical properties of its constituents.
>
> *Complexity.* Dembski defines the complexity of an object in terms of the probability of its appearance. An object that is highly improbable is by the same token highly complex.
>
> *Specification.* Some patterns look like gibberish; some do not. Dembski calls a pattern that does not look like gibberish *specified.* More precisely, if a pattern resembles a known target, then that pattern is specified. If it does not, then it is a *fabrication.*

Many of Dembski's examples involve coin tosses. He imagines flipping a coin many times and calls the resulting sequence of heads and tails a pattern. He calculates the probability of a given pattern by assuming he has an unbiased coin that gives the same probability of heads as of tails—that is, $1/2$. Using an argument based on the age of the universe, Dembski concludes that a contingent pattern that must be described by more than 500 bits of information cannot have formed by chance, although he is inconsistent about this limit in his examples.

If a pattern is both specified and complex, then it displays *specified complexity,* a term that Dembski uses interchangeably with *complex specified information.* Specified complexity, according to Dembski, cannot appear as the result of purely natural processes. Chapter 7 shows that specified complexity is inherently ill-defined and does not have the properties Dembski claims for it. Indeed, Dembski himself calculates the specified complexity of various events inconsistently, using one method when it suits him and another at other times.

In one example, he dismisses Bénard cells as examples of naturally occurring complexity; they form, he says, as a direct result of the properties of water. Chapter 7 shows that, to the contrary, Bénard cells are highly complex. Chapter 2 also shows how Dembski dismisses the formation of a complex entity such as a snowflake in the same way.

Dembski employs an explanatory filter that purports to use the concepts of contingency, complexity, and specification to distinguish design from chance and necessity. He argues that forensic scientists and archaeologists use a variation of the explanatory filter to infer design in those instances in which the designer is presumed to be human. Chapter 8 shows that forensic scientists do not solve problems using an explanatory filter; specified complexity and the explanatory filter do not provide a way to distinguish between designed objects and undesigned objects. Indeed, what Dembski calls side information is more important to a forensic scientist than the explanatory filter, which is virtually useless.

Attempts to distinguish rigorously between data that exhibit interesting patterns and data that are the result of simple natural laws or chance are not new. Chapter 9 explores approaches to this problem based on established complexity theory, a part of theoretical computer science. It shows that Dembski's idiosyncratic approach does not deliver what it promises and that mainstream science has much better ways to approach interesting questions about complex information.

Chapter 10 shows that randomness can help create innovation in a way that deterministic processes cannot. A hill-climbing algorithm that cannot see to the next hill may get stuck on a fairly low peak in a *fitness landscape*; further progress is thereby precluded. On the other hand, a random jump every now and then may well carry the algorithm to the base of a taller peak, which it can then scale. Randomness is not inimical to evolution; on the contrary, randomness is critical for its ability to produce genuine creative novelty. Chapter 10 draws upon artificial-intelligence research to show that intelligence itself may be explainable in terms of chance plus necessity, a combination that escapes Dembski's explanatory filter with its stark black-and-white dichotomies.

Dembski extends his argument by applying the no-free-lunch theorems (NFL theorems) to biological evolution. These theorems apply to computer-aided optimization programs that are used, for example, to design a lens by a series of trial-and-error calculations that begin with a very poor design. Roughly, an optimization program is like a strategy for finding the highest mountain in a given range; the height of the mountain represents the value of some figure of merit that we calculate as we go along and whose value we try to maximize.

The NFL theorems, according to Dembski, show that no search algorithm performs better than a random search. In fact, chapter 11 shows that the theorems are much more restricted than Dembski makes out; they state only that no strategy is better than any other when averaged over all possible mountain ranges, or fitness landscapes. In practice, however, we are almost never interested in all possible fitness landscapes but in very specific landscapes. It is entirely possible to design a strategy that will outperform a random search in many practical fitness landscapes. In addition, the NFL theorems apply only to landscapes that are fixed or vary independently of an evolving population, whereas the fitness landscape in biological evolution varies with time as organisms change both themselves and their environments. Thus, Dembski's application of the NFL theorems is wrong on two counts.

In cosmology, intelligent-design advocates point to the supposed fine tuning of the physical constants and claim that life would not exist if any of several physical constants had been slightly different from their present values—for example, because the lifetime of the universe will be too short for stars to form. Chapter 12 criticizes this anthropic argument, which suggests that the physical constants of our universe were purposefully designed to produce human life. The chapter notes, first, that the claim inherently assumes only one possible kind of life: ours. Additionally, this chapter shows that many combinations of values of four physical constants will lead to a universe with a long-enough life for stars to form and hence for life to be a possibility.

Chapter 13 asks whether, after all, intelligent design is practiced as science. To this end, it shows how certain pathological sciences operate and how they differ from genuine science. Specifically, we argue that the advocates of intelligent design do not practice science, not because their ideas are religiously motivated but because they make no substantive predictions, do not respond to evidence, have an ax to grind, and appear to be oblivious to criticism. Further, we hoist Dembski by his own petard when we demonstrate that his intelligent designer is no more than a Z-factor, a term of derision he applies to certain speculative scientific theories.

Chapter 1

Grand Themes,
Narrow Constituency

TANER EDIS

In the beginning, there was young-earth creationism. Even now, long after evolution has conquered the scientific world, "scientific" creationism remains popular, periodically surfacing to complicate the lives of science educators. This old-time creationism, however, has major shortcomings. Its religious motives are too obvious, its scientific credentials next to nonexistent. There is an aura of crankishness about claiming that special creation is not only scientific but also better than what ordinary science has to offer. In mainstream scientific circles, creationism produces exasperation and sometimes a kind of aesthetic fascination with the sheer extent of its badness. So scientists engage with creationists in a political struggle, not a serious intellectual dispute. Although they may miss opportunities to address some interesting questions (Edis 1998b), there is a limit to the excitement of continually revisiting matters resolved in the nineteenth century.

A new species of creationism, fighting evolution under the banner of intelligent design (ID), is attempting to change this picture. Many ID proponents not only sport Ph.D.s but have also done research in disciplines such as mathematics, philosophy, and even biology. They disavow overly sectarian claims, steering away from questions such as the literal truth of the Bible. And instead of trafficking in absurdities like flood geology, they emphasize grand intellectual themes: that complex order requires a designing intelligence, that mere chance and necessity fall short of accounting for our world (Moreland 1994, Dembski 1998a, Dembski 1999, Dembski and Kushiner 2001). They long to give real scientific teeth to intuitions about order and design shared by diverse philosophical and religious traditions.

At first, we might have expected ID to have a broad-based appeal. Scientists accustomed to evolution and wary of political battles over creationism might have been skeptical; but science is, after all, only one corner of intellectual life. Perhaps ID proponents could appeal to wider concerns and persuade scientists to reconsider intelligent design as an explanation for nature. At the least, it might spark an interesting debate about science and religion as ways of approaching our world and as influential institutions in society.

Curiously, though, very little of this debate has taken place. Academically, ID is invisible, except as a point of discussion in a few philosophy departments. Instead of treating it as a worthy if mistaken idea, scientists typically see it as the latest incarnation of bad, old-fashioned creationism. There has been little support for ID in nonscientific intellectual circles; even in academic theology, it has made inroads only among conservatives. ID promised to be broad-based but could not go beyond the old creationism's narrow constituency. It was supposed to be intellectually substantial, but scientists usually treat it as a nuisance. Most disappointingly, ID attracts attention only because it turns up in endless, repeated political battles over science education.

So what went wrong? Why has the intellectual response to ID ranged from tepid to hostile?

Design, East and West

Stepping outside the western debate over evolution may help us put ID into perspective. Islam has lately attracted much attention as a resurgent scripture-centered faith in a time of global religious revival. It appears to be an exception to the thesis that secularization is the inescapable destiny of modernizing societies, and it impresses scholars with the vitality of its religious politics. Less well known, however, is the fact that the Islamic world harbors what may be the strongest popular creationism in the world and that the homegrown intellectual culture in Muslim countries generally considers Darwinian evolution to be unacceptable.

In Turkey, which has felt modernizing pressures more than most Islamic countries, both a richly supported, politically well connected, popular creationism and a creationist influence in state-run education have appeared over the past few decades (Edis 1994, 1999; Sayin and Kence 1999). In Islamic bookstores from London to Istanbul, attractive books published under the name of Harun Yahya appear, promising everything from proof of the scientific collapse of evolution (Yahya 1997) to an exposition that Darwinism is fundamentally responsible for terrorist events such as that of 11 September 2001 (Yahya 2002).

Yahya's work is the Muslim equivalent of old-time creationism in the United States; indeed, it borrows freely from U.S. creationist literature, adapting it to a Muslim context by downplaying inessential aspects such as flood geology. In both its politics and its ability to reach beyond a conservative religious subculture, it is more successful than its U.S. counterpart.

Islamic creationism has much closer ties to intellectual high culture than in the United States. It would be nearly impossible for a creationist book to win endorsements from a prestigious U.S. divinity school, but Yahya's books print the praise of faculty members in leading Turkish departments of theology. One reason is that, in Muslim religious thought, the classical argument from design retains an importance it has long since lost in the west. Partly because of Quranic antecedents, Muslim apologetics at all levels of sophistication often rely on a sense that intelligent design is just plain obvious in the intricate complexities of nature (Edis 2003).

In other words, a kind of diffuse, taken-for-granted version of ID is part of a common Muslim intellectual background. The grand themes of ID are just as visible in the anti-evolutionary writings of Muslims who have more stature than Yahya. Osman Bakar (1987), vice-chancellor of the University of Malaya, criticizes evolutionary theory as a materialist philosophy that attempts to deny nature's manifest dependence on its creator and throws his support behind the endeavor to construct an alternative Islamic science, which would incorporate a traditional Muslim perspective into its basic assumptions about how nature should be studied (Bakar 1999). His desire is reminiscent of theistic science as expressed by some Christian philosophers with ID sympathies, which includes a built-in design perspective as an alternative to naturalistic science (Moreland 1994, Plantinga 1991). Seyyed Hossein Nasr (1989, 234–44), one of the best-known scholars of Islam in the field of religious studies, denounces Darwinian evolution as logically absurd and incompatible with the hierarchical view of reality that all genuine religious traditions demand, echoing the implicit ID theme that ours must be a top-down world in which lower levels of reality depend on higher, more spiritual levels.

The notion of intelligent design, as it appears in the Muslim world or in the western ID movement, is not just philosophical speculation about a divine activity that has receded to some sort of metaphysical ultimate. Neither is it a series of quibbles about the fossil record or biochemistry; indeed, ID's central concern is not really biology. The grand themes of ID center on the nature of intelligence and creativity.

In the top-down, hierarchical view of reality shared by ID proponents and most Muslim thinkers, intelligence must not be reducible to a natural phenomenon, explainable in conventional scientific terms. As John G. West, Jr., (2001) asserts:

> Intelligent design . . . suggests that mind precedes matter and that intelligence is an irreducible property just like matter. This opens the door to an effective alternative to materialistic reductionism. If intelligence itself is an irreducible property, then it is improper to try to reduce mind to matter. Mind can only be explained in terms of itself—like matter is explained in terms of itself. In short, intelligent design opens the door to a theory of a nonmaterial soul that can be defended within the bounds of science. (66)

Accordingly, ID attempts to establish design as a "fundamental mode of scientific explanation on a par with chance and necessity"—as with Aristotle's final causes (Dembski 2001b, 174).

Intelligence, of course, is manifested in creativity. ID proponents believe that the intricate, complex structures that excite our sense of wonder must be the signatures of creative intelligence. The meaningful information in the world must derive from intelligent sources. The efforts of mathematician and philosopher William Dembski (1998b, 1999), the leading theorist of ID, have been geared toward capturing this intuition that information must be something special, beyond chance and necessity.

The western ID movement has few Muslim connections. Among Muslims involved with ID, the most notable is Muzaffar Iqbal, a fellow of the International Society for Complexity, Information, and Design, a leading ID organization. Iqbal is also part of the Center for Islam and Science, a group of Muslim intellectuals promoting "Islamic science." But the connection is deeper than minimal organizational contact. The grand themes of ID resonate with a Muslim audience: they are found in much Muslim writing about evolution and how *manawi* (spiritual) reality creatively shapes the *maddi* (material). This is no surprise, because these themes are deeply rooted in any culture touched by near-eastern monotheism. They have not only popular appeal but the backing of sophisticated philosophical traditions developed over millennia.

Today, a full-blown defense of these themes must include a critique of modern biology. After all, while life, with its wondrous functional complexity, was once the poster child for the argument from design, it has now become the prime illustration of how to explain nature through chance and necessity. Evolution in the minimal sense of descent with modification could be accommodated if it could be seen as a progression toward higher orders of being; indeed, such was the initial response of even evangelical theologians to Darwin (Livingstone 1987). Interpreting evolution as an explicitly guided development would retain a sense of intelligent design; and this approach is still alive among more liberal thinkers, both Christian and Muslim. Darwinian biology, however, strains this view since it relies on nothing but blind mechanisms with no intrinsic directionality. The main sticking point is not descent with modification or progress but *mechanism*: chance and necessity suffice; hence, design as a fundamental principle disappears.

Defenders of intelligent design, then, understandably feel a need to pick a quarrel with Darwinian evolution. In the Muslim world, this task is more straightforward because a generic philosophical version of ID is part of the intellectual background. This is no longer the case in western intellectual life. The ID movement here is attempting to regain a foothold in the intellectual culture. To do so, proponents need to flesh out their intuitions about design and put them into play as scientific explanations. Thus, it is westerners, not Muslims, who invent notions of irreducible complexity in molecular biology (Behe 1996) and try to formulate mathematical tests to show that information is something special, beyond mere mechanisms, and a signature of design (Dembski 1998b).

ID among the Theologians

ID involves philosophy and theology, as well as attempts at science, and the grand themes it tries to defend might seem more at home in theology than in science. Indeed, the movement has attracted a number of philosophers and theologians with conservative religious commitments: Alvin Plantinga, Stephen C. Meyer, J. P. Moreland, William A. Dembski, William Lane Craig, Robert C. Koons, Jay Wesley Richards, John Mark Reynolds, Paul A. Nelson, Bruce L. Gordon, and no doubt many others (Moreland 1994, Dembski 1998a, Dembski and Kushiner 2001).

Academic theology in general, however, has a more liberal bent; it is not inclined to challenge mainstream science. Even so, we might expect some of ID's concerns and themes to surface in the west. After all, its central concerns do not involve minor sectarian points of doctrine but notions of divine design that should have a broad appeal.

Some echoes of ID's preoccupations can, in fact, be found in the writings of theological liberals who are friendly toward evolution. John F. Haught (2003), who vigorously defends the view that modern biology is fully compatible with Christianity and criticizes the ID movement for its theological lack of depth, nevertheless believes that creative novelty cannot be captured by mere mechanism, by chance and necessity. Like ID proponents, he takes information to be a key concept, describing God as "the ultimate source of the novel informational patterns available to evolution" (Haught 2000, 73). Another well-known example comes from the work of John Polkinghorne (1998) and Arthur Peacocke (1986), who speculate about how the indeterminism in modern physics might allow us to speak of a top-down sort of causality, beyond chance and necessity, which is connected to "active information" and allows intelligent guidance of evolution.

Curiously, academic theologians are often more willing to defend ID-like ideas outside the context of biological evolution. For example, some religious thinkers are enamored of parapsychology, which gets scarcely more respect than ID does in scientific circles. Accepting the reality of psychic powers, they see evidence that mind is independent of matter, that "agent causation" is an irreducible category of explanation very similar to design as ID proponents conceive of it (Stoeber and Meynell 1996).

It is notable, though, that such echoes of ID are merely echoes; only someone looking for parallels would notice them. These ideas seem to come up independently of the ID movement, appearing without favorable citation of any ID figure. Moreover, the echoes remain wholly undeveloped and tentative. For example, Polkinghorne never advances his speculations about information and quantum randomness as a space for divine action. Doing so would mean making the strong claim that the randomness in modern physics is not truly random and that a pattern might be revealed, perhaps brought to light by a design argument. Rather, he leaves his ideas at the "could be that" stage, never directly engaging science.

This brings up an intriguing possibility: that ID can be a means of bridging the gulf separating conservative and liberal theologies. Conservatives suffer from a reputation for intellectual backwardness, liberals from the impression that they are too accommodating, too given to compatible-with-anything hand waving. ID might provide conservatives with sophistication and liberals with a more-solid formulation for their intuition. This does not even necessitate a complete denial of evolution. After all, the grand themes of ID do not require that descent with modification be false, just that mere mechanisms not be up to the task of assembling functional complexity. Technically, Dembski's theories of ID do not require divine intervention all the time. The information revealed in evolution could have been injected into the universe through its initial conditions and then left to unfold (Edis 2001). So there is at least the possibility of some common ground.

But of course, liberals and conservatives have not come closer. The ID movement remains theologically conservative and harbors a deep distrust of descent with modification, not only of Darwinian mechanisms. Dembski (2002b, 212) has made a few half-hearted statements to the effect that even if modern biology remains intact, his work will show that an intelligent designer is the source of all genuine creativity. It is unlikely, however, that the ID movement will take this direction.

On their part, liberal religious thinkers about evolution usually do not treat ID as a religious option worth exploring. One exception is Warren A. Nord (1999), who has included ID among the intellectually substantive approaches he thinks biology education should acknowledge alongside a Darwinian view:

> Yes, religious liberals have accepted evolution pretty much from the
> time Charles Darwin proposed it, but in contrast to Darwin many of
> them believe that evolution is purposeful and that nature has a
> spiritual dimension. . . . Biology texts and the national science
> standards both ignore not only fundamentalist creationism but also
> those more liberal religious ways of interpreting evolution found in
> process theology, creation spirituality, intelligent-design theory and
> much feminist and postmodern theology. (712)

Such acknowledgment of ID is notably rare. It has more to do with Nord's (1995) long-standing insistence that more religion should be incorporated into public teaching than with his acceptance of ID in academic theology.

No doubt, this lack of contact largely reflects a cultural split. Liberal religion not only adapts to the modern world but is, in many ways, a driving force behind modernity. It has embraced modern intellectual life and ended up much better represented in academia than among the churchgoing public. By and large, it has been friendly to science, preferring to assert compatibility between science and a religious vision mainly concerned with moral progress. One result has been a theological climate in which the idea of direct divine intervention in the world, in the way that ID proponents envision, seems extremely distasteful.

The debate over ID easily falls into well-established patterns. ID arose from a conservative background, and conservatives remain its constituency. Its perceived attack on science triggers the accustomed political alignments already in place during the battle over old-fashioned creationism, when liberal theologians were the most reliable allies of mainstream science. What is at stake in this battle is not so much scientific theory as the success of rival political theologies and competing moral visions.

But if science is almost incidental to the larger cultural struggle, it is still crucial to how ID is perceived. In our culture, science enjoys a good deal of authority in describing the world; therefore, ID must present a scientific appearance. Although liberal religious thought has been influenced by postmodern fashions in the humanities and social sciences, resulting in some disillusionment with science, liberals still usually seek compatibility with science rather than confrontation.

So what scientists think of ID is most important for its prospects, more important than its fortunes in the world of philosophy and theology. ID has appealed only to a narrow intellectual constituency mainly because it thus far seems to be a scientific failure.

ID and the Scientists

The reaction of the scientific community to ID has been decidedly negative. Like many advocates of ideas out of the mainstream, ID proponents are given to suspect that their rejection has more to do with prejudice than with a fair consideration of merit. This suspicion is especially strong since ID has religious overtones, no matter how neutrally they come packaged. After all, it has long been conventional wisdom that science and religion have separate spheres and that scientists do not look kindly upon religious encroachment on their territory.

This is not to say that scientists are biased against religion. In fact, although there is considerable skepticism among scientific elites (Larson and Witham 1998), workers in scientific fields are not hugely different from the general population in their religious beliefs (Stark and Finke 2000, 52–55). Nevertheless, there may be institutional barriers to the fair consideration of scientific claims with religious connotations.

Such suspicions within the ID movement are reinforced when the first defense of evolution they encounter is that their ideas are intrinsically unscientific—that science cannot even properly consider non-naturalistic claims such as ID, let alone accept them. Therefore, much of the philosophical effort behind ID has been devoted to defeating this presumption of methodological naturalism (Pennock 1996). Reading methodological naturalism as a strict requirement for doing science is, in fact, overly strong. The philosophy of science is littered with failed attempts to define an essence of science, separating legitimate hypotheses from those that fall beyond the pale. At any one time, a list of such requirements—naturalism, repeatability, and so on—might appear plausible. If so, it is because they are abstracted from successful practice, not because they are inevitable requirements of some disembodied Reason. Such principles may even inspire a research program, but like behaviorism in psychology, which countenanced only the directly observable, they can fail.

Confining science to naturalistic hypotheses would also be historically strange. Biologists of Darwin's day, for example, compared evolution to special creation as rival explanations and argued that evolution was superior, not that creation should never have been brought up. Even today, explanations in terms of the intentions and designs of persons are legitimately part of historical or archaeological work. Today's state of knowledge might incline us to think such agent-causation is eventually reducible to chance and necessity, but we need not assume this is so in order to do science.

ID philosophers bring up many such objections, and they are largely correct. Methodological naturalism cannot be used as an ID-stopper. If it is to fail, ID should be allowed to fail as a scientific proposal. On the other hand, naturalism may still make sense as a methodology, justified not by philosophical fiat but by historical experience.

Consider an astrophysicist studying distant galaxies. She will, in constructing her theories, assume that physics is the same out there as it is here: that the same sort of particles interact in the same way we observe them to do close to home, that gravity does not suddenly act by an inverse-cube law outside our galaxy. This does not mean that the only legitimate astrophysical hypotheses follow this assumption. After all, in certain ways, such as the presence of life, our corner of the universe may well be unrepresentative. Not too many centuries ago, our physics was Aristotelian: the sublunar realm was supposed to behave in ways radically different from what took place in the spheres beyond the moon. Assuming the same physics throughout the universe, however, has been successful in recent history, and no current rivals promise better explanations. Assuming that physics is the same is our best bet, likely keeping us from wasting time on fruitless research. Similarly, preferring naturalistic theories makes the best sense in light of our successful experience with theories such as evolution (Richter 2002).

This does not mean that ID is disallowed. It means that ID is a very ambitious claim and that it must produce strong evidence before scientists go along with the proposed revolution. Success for ID should be difficult, but not out of reach.

Is the scientific community open to such evidence? The answer has to be a qualified yes. Scientists are often conservative, resistant to changing their theories; practical methodologies may well harden into blinders over time. But scientists also need new ideas to advance their work, and they do not pay much attention to the lists that philosophers make to define science. Even if methodological naturalism is the reigning conventional wisdom, it is not absolute dogma, and ID can still reach a scientific audience.

One important way for unorthodox ideas to gain a hearing is through scientific criticism. It does not greatly matter if the critics are initially hostile. To avoid embarrassment, if for no other reason, critics must at least understand the unfamiliar ideas and learn to work with them. Otherwise, an adequate job of criticism will not be possible. This learning process has historically been important in the acceptance of many revolutionary views, including Darwinian evolution itself (Thagard 1992). Critics can become converts.

Another way might be for a few scientists, perhaps those who are young and less committed to evolution than their elders, to take their chances with ID. If they can succeed with research driven by an ID perspective, consistently producing results that are surprising from an evolutionary standpoint, ID will suddenly be taken much more seriously.

But ID does not seem to be moving forward at all in the scientific world. It does not lack serious critics who are willing to engage with its claims in technical detail. Far from being converted, the critics consistently find ID's claims to be disappointing. Its most significant biological effort has been Michael Behe's argument for irreducible complexity, which turned out to be very poor work, not to mention current progress on the very problems Behe had said were not being addressed from a Darwinian viewpoint and could not be (Miller 1999, Shanks and Joplin 1999). William Dembski, ID's wunderkind in information theory, produced work that might eventually contribute to detecting an interesting type of complex order, but it has no bearing on the truth of Darwinian evolution (Edis 2001). Since then, Dembski has been busy misapplying certain mathematical ideas to prove that the Darwinian mechanism cannot be truly creative (Rosenhouse 2002).

The young Turks who might do novel research based on ID also have not materialized. This is not to say the biology departments of American universities are devoid of the occasional faculty member with ID sympathies. Not a few must have prior religious commitments that incline them toward ID. But productive, surprising research *driven by ID* is noticeably absent.

ID might one day make its big push. Perhaps it is too early, and ID's research ideas have not been fully developed yet. Perhaps. But so far ID has been singularly unproductive, and nothing about it inspires confidence that things will change. It is no wonder that ID gets no respect from the scientific community.

Politics, Again

With its ambitions to be the intellectually sophisticated opposition to Darwinian evolution, ID has failed to make headway among intellectual elites. But it has the solid support of a popular religious movement, the same constituency that supported old-fashioned creationism. Understandably, ID proponents have been trying to play to their strength. The movement today looks more like an interest group trying to find political muscle than a group of intellectuals defending a minority opinion. Like their creationist ancestors, they continually make demands on education policy. Similarly, their arguments against evolution do not build a coherent alternative view but collect alleged "failures of Darwinism."

Unfortunately for ID, there is no crisis in Darwinian evolution. Its vitality can be judged best by observing not only its nearly universal acceptance in biology but the way in which Darwinian thinking has come to influence other disciplines. From speculations in physical cosmology (Smolin 1997) to influential hypotheses in our contemporary sciences of the mind, variation-and-selection arguments have come to bear on many examples of complex order in our world. To some, this suggests a universal Darwinism that undermines all top-down, spiritual descriptions of our world (Dennett 1995, Edis 2002), while others argue that the Darwinian view of life is no threat to liberal religion (Ruse 2001, Rolston 1999).

ID, however, is not part of this debate. Darwinian ideas spilling out of biology can only confirm the suspicions of ID proponents that Darwinism is not just innocent science but a materialist philosophy out to erase all perceptions of direct divine action from our intellectual culture. So they have plenty of motivation to continue the good fight. In the immediate future, however, the fight will not primarily involve scientific debate or even a wider philosophical discussion but an ugly political struggle.

Chapter 2

Grand Designs and
Facile Analogies

Exposing Behe's Mousetrap
and Dembski's Arrow

MATT YOUNG

*Though analogy is often misleading, it is often the least
misleading thing we have.*

—Samuel Butler, *Notebooks*, 1912

Mᴜᴄʜ ᴏꜰ ᴡʜᴀᴛ we know or think we know, it seems to me, is based on analogy. When we describe a gas as a collection of colliding billiard balls, our model is based on an analogy. When we think of a gene as fighting for its survival against other genes, our model is based on an analogy. When we describe a photon as a wave or a particle, however, our analogy breaks down, because the photon has both wavelike and particle-like properties.

It is thus necessary to use analogy judiciously.

The neocreationists Michael Behe (1996) and William Dembski (1999, 2002b) do no such thing with their analogies of the mousetrap and the archer (Perakh 2001b). Behe, in particular, expects his analogy to bear the heavy burden of illustrating his point. If the analogy fails, then the entire argument is likely to fail. Dembski, likewise, leans heavily on a flawed analogy and covers up its failure with abstruse mathematical notation and invented jargon.

Behe's Opaque Box

Behe, a biochemist, argues that his own field is somehow more fundamental than all others. He notes that biochemistry is extremely complicated and points to some systems that are so complicated that they are irreducibly complex.

IRREDUCIBLE COMPLEXITY

Behe says, in essence, that a system is irreducibly complex if it includes three or more parts that are crucial to its operation. The system may have many more than three parts, but at least three of those parts are crucial. An irreducibly complex system will not just function poorly without one of its parts; it will not function at all.

Behe has found several biochemical systems that he claims are irreducibly complex. Such systems, he argues, cannot have evolved gradually by a series of slight modifications of simpler systems, because they will not work at all if one of their crucial parts is missing. I do not want to discuss Behe's claim in detail (for that, see chapters 4, 5, and 6); rather, I want to concentrate on his analogy of the mousetrap.

Behe uses the common mousetrap to exemplify a system that, he claims, is irreducibly complex. Figure 2.1 shows a mousetrap, which includes a hammer, a spring, a pin (which passes through the center of the spring), a latch, a bar, a platform, and a handful of other parts such as staples. The bait, which is not crucial to the operation of the trap, is not shown. The bar is used to hold the hammer in place and is in turn held in place by the latch. When the mouse takes the bait, he dislodges the bar and frees the hammer, which is driven by the spring and snaps closed, with unfortunate consequences for the mouse.

Behe claims that the mousetrap is irreducibly complex—that is, that it cannot function without all of its parts. The statement is entirely wrong. I have acquired a mousetrap and removed the latch. In the mousetrap I used,

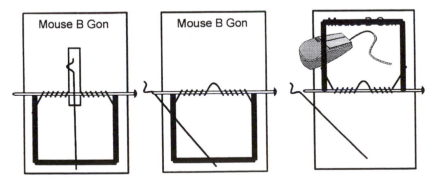

Figure 2.1. Left, a mousetrap with the conventional latching mechanism. Center, with the latch removed. Right, the mousetrap still functions.

it was a simple matter to wedge the bar under the pin in such a way that it was barely stable, as shown in figure 2.1, center. No bending or filing was necessary. If the mouse dislodged the bar from the left side of the trap, he probably got away with a good scare. But if he dislodged the bar from the right (platform) side, he was probably caught. Thus, the reduced mousetrap is not nearly as good as the entire trap, but it still works.

My mousetrap has seven parts, not counting staples. Could Behe merely have counted wrong, and is the mousetrap with six parts irreducibly complex? No. John McDonald (2000) has shown that the trap can be reduced successively to fewer and fewer parts until only the spring remains. Dembski (2002b) has criticized McDonald's approach, claiming that McDonald modified some parts as he removed others.

Irreducible complexity, to Dembski, means that a given part has to be removed with no changes to any of the others. And that points out exactly what is wrong with the concept of irreducible complexity.

In biology, parts that were used for one purpose may be co-opted and used for another purpose. A well-known example is the development of the mammalian ear from reptilian jaw bones. Specifically, Stephen Jay Gould (1993) gives good evidence that bones that originally supported the gills of a fish evolved first into a brace for holding the jaw to the skull and later into the bones in the inner ear of mammals.

Those jaw bones did not just suddenly one day re-form themselves and decide to become ear bones; because mammals did not need unhinging jaws, the jaw bones gradually changed their shape and their function until they became the bones of the inner ear. The ear today may be irreducibly complex, but once it was not. You might, however, be fooled into thinking that the ear could not have evolved if you did not know exactly how it originated.

BLUEPRINTS VERSUS RECIPES

What is the difference between a mouse and a mousetrap? Or, more precisely, how do mice propagate, and how do mousetraps propagate (Young 2001a)?

Mousetraps are not living organisms. They do not multiply and do not carry within them the information necessary to propagate themselves. Instead, human engineers propagate them by blueprints, or exact specifications (see table 2.1). Each mousetrap in a given generation is thus nominally identical to every other mousetrap. If there are differences among the mousetraps, they are usually not functional, and they do not propagate to the next generation. Changes from one generation to the next, however, may be very significant, as when the designer triples the strength of the spring or doubles the size of the trap and calls it a rattrap.

Table 2.1
How mice differ from mousetraps

Mousetraps Propagate by Blueprints	*Mice Propagate by Recipes, not Blueprints*
Exact specifications	*Inexact* specifications
Errors do not propagate	Errors propagate
Finite changes, if any	*Small* or infinitesimal changes
All mousetraps of given generation identical	All mice of given generation different

The genome, by contrast, is a recipe, not a blueprint. The genome tells the mouse to have hair, for example, but it does not specify where each hair is located, precisely as a recipe tells a cake to have bubbles but does not specify the location of each bubble. Additionally, the specifications of each mouse differ from those of each other mouse because they have slightly different genomes. Thus, we could expect a mouse to evolve from a protomouse by a succession of small changes, whereas we can never expect a mousetrap to evolve from a prototrap.

This is so because the mousetrap is specified by a blueprint, the mouse by a recipe. If improvements are made to a mousetrap, they need not be small. It is therefore no criticism of McDonald to argue, as Dembski does, that McDonald cannot reverse-engineer a complex mousetrap by building it up from his simpler examples. It is Behe, not McDonald, who has erred in using the mousetrap as an analog of an evolving organism, precisely because the parts of an evolving system change as the system evolves.

HALF A FLAGELLUM

Behe argues that an irreducibly complex system cannot evolve by small changes. His preferred example is the flagellum, and he asks, in essence, "What good is half a flagellum?" A flagellum without its whiplike tail or without its power source or its bearing cannot work. Behe cannot imagine how each part could have evolved in concert with the others, so he decides it could not have happened. In this respect, he echoes the smug self-confidence of the creationist who asks, "What good is half an eye?" An eye, according to the creationist, has to be perfect or it has no value whatsoever.

This logic is easily debunked (Young 2001b, 59–62, 122–23). As any nearsighted person will tell you, an eye does not have to be perfect in order to have value. An eye does not even have to project an image to have value. Indeed, the simplest eye, a light-sensitive spot, gives a primitive creature warning of an approaching predator.

Biologists Dan Nilsson and Susanne Pelger (1994) have performed a sophisticated calculation to show that an eye capable of casting an image could
evolve gradually, possibly within a few hundred thousand years, from a simple
eye spot, through a somewhat directional eye pit, to a spherical eye that cannot change focus, and finally to an eye complete with a cornea and a lens.
Because the eye is composed of soft tissue, we do not have fossil evidence of
the evolution of eyes in this way. Nevertheless, every step that appears in
the calculation is represented in some animal known today. The inference that
eyes evolved roughly as suggested in the calculation is therefore supported
by hard evidence. (See Berlinski [2002, 2003] and Berlinski and his critics
[2003a, 2003b] for a surprisingly intemperate attack against Nilsson and Pelger's
paper.)

The eye is not irreducibly complex. You can take away the lens or the
cones, for example, and still have useful if impaired vision. Nevertheless, it
was used for years as an example of a system that was too complicated to have
evolved gradually. It is not, and neither is the eukaryotic flagellum (Stevens
1998, Cavalier-Smith 1997) or the bacterial flagellum (see chapters 4 and 5).

EMERGENCE

The physical world can be thought of as a series of levels, each underlain by
a lower level but largely isolated from that level. Thus, the viscosity of water
can be explained in terms of molecular physics, but you do not have to understand molecular physics to appreciate viscosity and indeed to study it. Viscosity is an example of an *emergent property*, a property that, in this case,
appears only when we assemble a large number of water molecules under certain conditions of pressure and temperature.

Emergent properties are the result of self-organization (see chapter 7) and
force reality into a series of levels: biochemical, organelle (an "organ" within
a cell), cell, organ, organism, . . . for example. No one level is more fundamental
than any other. Liquid water is no more fundamental than isolated water molecules are. Rather, each level is an alternate way of looking at reality. Viscosity does you no good if you are interested in the spectroscopy of water vapor,
whereas spectroscopy does you no good if you are interested in viscosity.

Creationism has failed at the level of the organism. We understand in
enough detail how an eye might have evolved to say with certainty that the
creationist's argument is no longer cogent. That argument is sometimes called
a God-of-the-gaps argument: a gap in our understanding is seen as evidence
for a divine creator.

Aware that the half-an-eye argument has failed, Behe has developed the half-a-flagellum argument. He has dressed it up with a rigorous-sounding term: irreducible complexity. But it is still the half-an-eye argument. Terminology aside, Behe's argument is pure God-of-the-gaps. According to him, if we do not know today how a flagellum could have evolved from simpler systems, then we never will. Chapter 6 shows, to the contrary, that we know a great deal more than Behe admits. Whenever we learn the evolution of the flagellum in enough detail, we may expect the Behe of that day to slither down another level and find an argument at the level of, say, physics, rather than biochemistry.

All that based on the flawed analogy of the mousetrap.

Dembski's Arrow

William Dembski (1999) invites us to consider an archer who shoots at a target on a wall. If the archer shoots randomly and then paints a target around every arrow, he says, we may infer nothing about the targets or the archer. On the other hand, if the archer consistently hits a target that is already in place, we may infer that he is a good archer. His hitting the target represents what Dembski calls a pattern, and we may infer design in the sense that the arrow is purposefully, not accidentally, centered in the target.

Using the archer as an analogy, Dembski notes that biologists find genes that are highly improbable yet not exactly arbitrary, not entirely gibberish. That is, the genes contain information; their bases are not arranged arbitrarily. Dembski calls such improbable but nonrandom genes complex because they are improbable and specified because they are not random (see chapter 9). A gene or other entity that is specified and complex enough displays specified complexity. Arrows sticking out of targets that have been painted around them are complex but not specified; arrows sticking out of a target that has been placed in advance are specified.

According to Dembski, natural processes cannot evolve information in excess of a certain number of bits—that is, cannot evolve specified complexity. His claim is not correct. We can easily see how specified complexity can be derived by purely natural means—for example, by genes duplicating and subsequently diverging or by organisms incorporating the genes of other organisms. In either case, an organism whose genome has less than the putative upper limit, 500 bits, can in a single stroke exceed that limit, as when an organism with a 400-bit genome incorporates another with a 300-bit genome (Young 2002). Here I want to concentrate not on the 500-bit limit but on the arrow analogy and its pitfalls.

MANY TARGETS

Consider a biological compound such as chlorophyll. Chlorophyll provides energy to plant cells, and most (but not all) of life on earth either directly or indirectly depends for its existence on chlorophyll. The gene that codes for chlorophyll has a certain number N of bits of information. Dembski would calculate the probability of that gene's assembling itself by assuming that each bit has a 50-percent probability of being either 0 or 1 (Wein 2002a). As I noted in connection with a book by Gerald Schroeder (Young 1998), such calculations are flawed by the assumption of independent probabilities—that is, by the assumption that each bit is independent of each other bit. Additionally, they assume that the gene in question has a fixed length and that the information in the gene has been selected by random sampling, whereas most biologists would argue that the gene developed over time from less-complex genes.

But Dembski makes a more-fundamental error: he calculates the probability of occurrence of a specific gene (T-urf13) and also considers genes that are homologous with that gene. In other words, he calculates the probability of a specific gene and only those genes that are closely related to that gene. In terms of the archer analogy, Dembski is saying that the target is not a point but is a little fuzzy. Nevertheless, calculating the probability of a specific gene or genes is the wrong calculation, and the error is exemplified in Dembski's archer analogy. (He makes another interesting conceptual error: On page 292 of *No Free Lunch*, Dembski (2002b) calculates the probability that all the proteins in the bacterial flagellum will come together "in one spot." Besides the assumptions of equal and independent probabilities, that is simply the wrong calculation. He should have calculated the probability of the genes that code for the flagellum, not of the flagellum itself. He treats the protein URF13 similarly on pages 218–19.)

Let us do a Dembski-style analysis using the example of chlorophyll. According to the *Encyclopedia Britannica*, there are at least five different kinds of chlorophyll. There may be potentially many more that have never evolved. Thus, the archer is not shooting at a single, specific target on the wall but at a wall that may contain a very large number of targets, any one of which will count as a bull's-eye. Dembski should have considered the probability that the archer would have hit any one of a great number of targets, not just one target.

Chlorophyll, moreover, is not necessary for life. We know of bacteria that derive energy from the sun but use bacteriorhodopsin in place of chlorophyll. Other bacteria derive their energy from chemosynthesis rather than photosynthesis. If we are interested in knowing whether life was designed, then we have to calculate the probability that any energy-supplying mechanism will

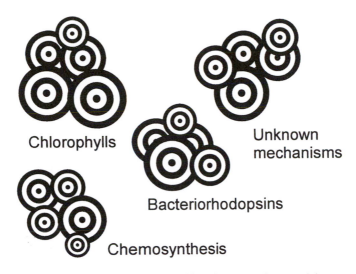

Figure 2.2. What Dembski's wall really looks like, showing only part of the vast array of possibilities for generating energy in plants.

evolve, not just chlorophyll. Thus, photosynthesis, chemosynthesis, and all their variants must show up as targets on the wall, as well as other perhaps wholly unknown mechanisms for providing energy to a cell, as suggested in figure 2.2. Additionally, we cannot rule out the possibility that there are other universes besides our own; and these, too, must be included in the calculation (see chapter 12).

I do not think Dembski is arguing that life takes a single shot at a target and either hits it or not; he knows very well that complexity was not born instantaneously. The target is a distant target, and the path is tortuous. But by using his archer analogy, Dembski implies that life is very improbable and the target impossible to hit by accident. It may or may not be: there are more galaxies in the known universe than there are stars in our galaxy. Life has arguably had a great many opportunities to evolve. That it evolved complexity *here* is no doubt improbable; that it evolved complexity *somewhere* is very possibly not. Dembski has, in effect, calculated the probability that a certain woman in New Jersey will win the lottery twice, whereas it is more meaningful to calculate the (much-higher) probability that someone, somewhere will win the lottery twice.

In terms that Dembski knows very well, his rejection region should have included many more possibilities than just a handful of homologous genes. In my example, the rejection region should have included a target for chlorophyll, a target for bacteriorhodopsin, a target for chemosynthesis, and so on.

Now, it is entirely possible that even such an extended rejection region (on every planet in every universe) will yield a very low net probability, but Dembski has shown no such thing. And he cannot since we do not know just how much of the wall is covered with targets, nor how many arrows the archer has launched to get just one hit, nor how many archers there are in the universe, nor even how many universes there are.

It is peculiar that Dembski makes this mistake because, when it suits him, he recognizes that you have to consider an ensemble of possibilities. Thus, in *No Free Lunch* (2002b, 221), he describes a calculation for designing a radio antenna that radiates uniformly in all directions. According to him, such an antenna can be designed by using a genetic algorithm: a mathematical formalism that gradually modifies the antenna, one step at a time, until the antenna radiates uniformly or nearly so. Oddly, the resulting antenna is not a regular geometric shape such as a pyramid but a tangled mass of wires.

Such an antenna, says Dembski, is highly improbable since it is one of an infinity of possible antennas. High improbability or low probability is by Dembski's definition complex. A genetic algorithm mimics biological evolution. If the genetic algorithm can generate complexity, then so can evolution by natural selection.

FIGURE OF MERIT

Dembski does not deny that the formula for describing the antenna exhibits specified complexity but claims instead that the specified complexity has been sneaked in. How?

The function that describes the radiation by the antenna in any direction is called the antenna pattern. The desired antenna pattern is uniform; that is, it can be described by a single constant value that represents the strength of the radiation in any direction. On a three-dimensional graph, such an antenna pattern is a sphere.

A real antenna pattern will differ measurably from a sphere. To quantify the difference between the real antenna pattern and the sphere, we define a fitness function (Kauffman 1995). Dembski claims that the engineers sneaked complexity into their calculation when they defined the fitness function (see chapter 11).

My job is to make hard subjects easy, rather than the other way around. Let us therefore consider a simple case: a string of three numbers, each of which can be either 0 or 1. Let us say that the string is most fit when all three numbers are 1. Initially, however, the three numbers are chosen randomly and are not all equal to 1. We want to use a mathematical algorithm (it does not matter which one) that repeatedly changes one or more of the three numbers in some random fashion until they converge on the fittest configuration (111).

For this purpose, we define an ad hoc fitness function: the sum of the three numbers. The fitness function thus takes the value 0 if all three numbers are equal to 0, 1 if one of the numbers is 1, 2 if any two of the numbers are 1, and 3 if all three of the numbers are 1. We apply our algorithm time after time until all the numbers are equal to 1—that is, until the value of the fitness function is 3.

Thus, we begin with a random configuration of the three numbers, say (010). The value of the fitness function is 1. We roll some dice or toss a coin to tell us how to rearrange the numbers according to some rule. If the fitness function becomes 2, we keep the new configuration, say (110); otherwise, we keep the old configuration and roll the dice again. We keep rolling until we attain the configuration (111).

The fitness function, then, is not some information that we sneaked in from the outside. It is not a look-up table that we import in its entirety. It is, rather, a series of numbers that we calculate as we go along. This series is derived from the values of the three numbers in the string and on nothing else. Indeed, we need not ever plot the entire fitness function, and we do not need to retain any but the previous value.

The preceding example is for illustration only. Those who take it too literally will argue that the search is targeted and that no new information was therefore generated. It is, however, easy to generalize to an untargeted search (see chapter 11). For example, we may increase the number of digits to a very large number and toss coins for a finite time or for a finite number of steps. The number of digits may be made to vary with time, perhaps even to covary with the evolution of the string. Such a search is untargeted, and we will get a different result each time we carry out a search.

Suppose that we wanted to manufacture a bow that could launch an arrow as far as possible yet cost as little as possible. How would we decide when the cost was too high for a given range? We might define a figure of merit that is equal to the range divided by the cost. Then if one design had a lower cost or a higher range, the figure of merit would increase. Thus, our job is to design a bow with the highest figure of merit. The figure of merit is just a number that we calculate as an aid to evaluating our success. The fitness function is all possible values of the figure of merit, plotted as a function of cost and range.

Far from importing the fitness function and thereby sneaking complexity into the problem, the electrical engineers repetitively calculated a figure of merit and kept recalculating it until it converged to the desired value. At most, they imported a relatively small number of values of the figure of merit, and they could have discarded all but the largest as they went along. In a sense, they imported a small amount of information from the fitness function as they carried out their calculation. The term *fitness function* can be misleading, however: no one would think that the entire fitness function had been imported if it had been called a sequence of figures of merit.

OF SPHERES AND FLAKES

Dembski thus agrees that the antenna is complex, but he argues that the genetic algorithm did not generate that complexity; rather, the engineers sneaked in complexity by choosing their fitness function judiciously and simply rearranged the information—that is, transferred it from the fitness function to the antenna. As we have seen, however, the fitness function is just a series of calculated values of a figure of merit and does not have to be imported wholesale.

On page 12 of *No Free Lunch*, by contrast, Dembski (2002b) tells us that a regular geometric pattern such as a snowflake is not complex because it has been formed "simply in virtue of the properties of water." As Mark Perakh (2002a) points out, the formation of a snowflake is by no means assured when a droplet of water crystallizes in the air. Under certain conditions, triangular or hexagonal crystals may form instead. The formation of such crystals depends on the chance occurrence of unusual weather conditions. Thus, the formation of a snowflake, though likely, depends on both chance and law, not just law. Further, each snowflake is slightly different from all others, and that difference depends on the continuously changing conditions of temperature and humidity that the snowflake experienced as it fell through the atmosphere. The formation of a particular snowflake is far from determined in the way that, say, the formation of a single crystal of salt is determined. (See chapter 7 for a related discussion of Bénard cells.)

A snowflake is only one of an infinity of possible patterns, so by Dembski's definition, any given snowflake ought to be every bit as complex as an antenna. But when it suits Dembski, a spherical pattern is complex, whereas a snowflake remains simple because it forms by necessity. The inconsistency is transparent.

A Tangled Web

Behe and Dembski have an agenda: to *prove* that an intelligence guides evolution rather than *find out* whether an intelligence does so (see chapter 13).

They have thus constructed a tangled web of analogies to convince impressionable readers that evolution must necessarily have been directed. I leave it to others to show that their chemistry and their mathematics are faulty. Here I point out only that their analogies themselves are faulty: the mousetrap is not irreducibly complex and did not evolve gradually, whereas irreducibly complex structures need not have been created out of whole cloth but could have evolved, like the mammalian ear, from borrowed components. Dembski's archer, meantime, is shooting at a wall that may for all we know be so thickly covered with targets that he will certainly hit several after a large number of shots.

Acknowledgments

Many thanks to Mark Perakh, Taner Edis, Gert Korthof, and Jason Rosenhouse for their critical reading of this chapter in manuscript.

Chapter 3

Common Descent

It's All or Nothing

GERT KORTHOF

If the living world has not arisen from common ancestors by
means of an evolutionary process, then the fundamental
unity of living things is a hoax and their diversity, a joke.
—Theodosius Dobzhansky (1964)

I WROTE THIS CHAPTER for everyone who believes that it is possible to accept something less than full common descent of all life. In my view, anything less than full common descent leads to both an arbitrary fragmentation of the tree of life and a logically inconsistent theory of descent and also conflicts with the evidence.

Carl Linnaeus, the Swedish botanist who invented the biological nomenclature still in use today, was a creationist. According to Linnaeus, "We count as many species as different forms were created in the beginning" (Mayr 1982, 258). Linnaeus's work and thinking were based on the concept of design by God. Species were fixed, and their systematic relationships reflected the divine plan. Despite his belief in the fixity of species, Linnaeus came to accept varieties within species. These varieties, he thought, resulted from changed conditions. By 1756, at the end of his life, he had concluded that the number of species within a genus might increase. This was not evolution as we know it today; Linnaeus suggested that in the original creation God formed only a single species as the foundation of each genus and left the multiplication of species within genera to a natural process of hybridization (Bowler 1989, 67).

French naturalist Georges-Louis Buffon had defended the fixity of species early in his career, but in 1766, nearly a hundred years before the publi-

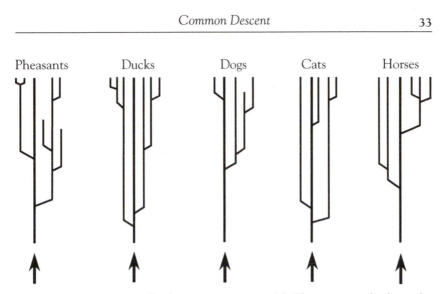

Pheasants Ducks Dogs Cats Horses

Figure 3.1. Basic types in the dynamic creation model. The arrows in the figure depict the creation of various "ground types." After Pennock (2001, 685).

cation of Charles Darwin's (1859) *Origin of Species*, he accepted the idea that closely related Linnaean species had diverged from a common ancestor. This view is close to what today is called microevolution. Buffon even went so far as to claim that families were created by God. The family possessed fixed characteristics and had no ancestors itself (Bowler 1989, 74).

Today, philosopher Paul Nelson (2001a, 684), who is part of the intelligent-design movement, argues for the creation of basic types (also known as "ground types") stemming from common ancestors. He illustrates those basic types with a figure similar to figure 3.1 (Junker and Scherer 1988). The illustration shows five animal groups: pheasants, ducks, dogs, cats, and horses. Each group is descended from a created common ancestor, which itself has no ancestor. Nelson contrasts a static-creation theory (creation of fixed species) with the dynamic-creation model he favors.

Nelson criticizes Mark Ridley (1985) for displaying only the static fixed-species model and ignoring the modern dynamic-creation model (see figure 3.2). The four publications Nelson uses as evidence for his accusation, including the source of figure 3.1 (Junker and Scherer 1988), postdate Ridley's 1985 book. His criticism is unfair because Ridley could not have known of publications appearing after his own book.

Nelson is not the only creationist proposing this kind of model. Jonathan Sarfati (2000, 38, 39) has a similar model, which he calls "the true creationist orchard." (Its picture looks like an orchard.) According to the model, diversity has occurred within the original Genesis kinds. There are no names

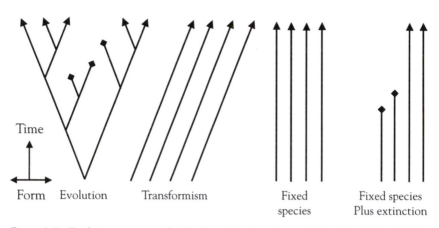

Figure 3.2. Evolution contrasted with three fixed-species models. After Ridley (1985, 1996).

attached to the trees in his illustration, so we can draw no inference as to their taxonomic level. He, too. contrasts his model with the alleged creationist "lawn," where each species is like a blade of grass, separate from all others.

The fixed-species model claims that no new species have been produced since creation, but it allows for extinction. The model is a caricature of creationism, says Sarfati, because it implies that the Genesis kinds were the same as today's species. He claims that the fixed-species model is a straw man for creationism, and Nelson claims that it has not appeared in creationist publications in recent decades. His statement implies that creationists adhered to the fixed-species model in previous decades. It would be interesting to know what insight or fact caused creationists to convert to the dynamic-creation model.

The dynamic-creation model is similar to the model of the creationist Walter Remine (1993). He places, for example, dogs, wolves, coyotes, jackals, and foxes in one systematic group: the *Canidae-monobaramin,* which matches Nelson's dogs group. Remine defines *monobaramin* as "a group containing only organisms related by common descent, sharing a common ancestor" (444). Subsequently, he states the inevitable: "Directly created organisms have no ancestor, they are created by the direct action of a designer" (510).

Implications of the Dynamic-Creation Model

The dynamic-creation model (DCM) uses the theological concept of creation as its foundation: "Here, the terminal species are members of basic types, stemming from common ancestors which were themselves created" (Nelson 2001a,

684). Therefore, DCM is basically theology. Nelson could have omitted the word *created*. He could have used an agnostic formula such as "the common ancestors of families are unknown and cannot be known to science," but he did not. Doing so would have destroyed the beating heart of creation theory.

In this chapter I ignore DCM's theological foundation not because it is a minor detail but because I want to avoid endless discussions about whether or not supernatural interventions are a legitimate part of natural science. I will first explore the biological implications of DCM and then evaluate the model itself.

1. *Implications for the taxonomic level of basic types.* Although Nelson does not state the taxonomic level of the groups, Siegfried Scherer (1998), the author of the drawing Nelson (2001a) uses, says that the basic types are families: Phasianidae (pheasants), Anatidae (ducks), Canidae (dogs), Felidae (cats), and Equidae (horses). Each basic type contains different genera (for example dog, fox, wolf), and each genus contains one or more species. Thus, families are created, not genera or species.

2. *Implications for the number of species originating from basic types.* The mini-trees in figure 3.1 show only five to eight species per tree. This misleads the reader: each genus has many more species than are depicted in the figure. The pheasant family consists of 38 genera and 155 species, the duck family 41 genera and 147 species, the dog family 12 genera and 34 species, and the cat family 37 species.

    These numbers are tiny in comparison with families of insects. A single beetle family (weevils) contains approximately 65,000 species (Tudge 2000). The supposition that the basic type contains all the information necessary to create all the descendant species is therefore highly implausible. It implies, for example, that the information for 65,000 weevil species was already present in the weevil basic type.

3. *Implications for the number of basic types.* How many basic types are there? Creationists don't tell us. Until we know, the dynamic-creation model is only a fragment of a theory. If basic types are to capture the million or so species on earth, the model must include thousands of basic types.

4. *Implications for the number of interventions.* The model includes only one arrow per basic type and thus one supernatural intervention per created basic type. The implication is that the rest of each mini-tree is free of supernatural interventions. If the author acknowledged the existence of additional interventions, he could no longer claim that creation had been at the family level. Indeed, such additional interventions might end up as the special creation of species, part of the static-creation

model, which Nelson emphatically rejects. Intelligent-design theorists do not deny the existence of unguided natural processes but claim that not all processes in nature are unguided. Therefore, they do not necessarily object to the idea that the mini-trees are unbroken chains of natural processes.

5. *Implications for mutation and natural selection.* If creation took place at the family level, then genera and species must have originated in a natural way. How? Since DCM claims that the basic types vary (within boundaries) by microevolutionary processes, the mechanism must be the standard Darwinian mutation and natural selection. These mechanisms produce all the terminal species. In other words, genera and species are created by natural processes as described in the textbooks.

   But then it does not make sense to keep talking about guided mutations and the like. Further, it does not make sense to continue objecting to the efficacy of natural selection—to claim that natural selection is not a creative force or that natural selection is a tautology and explains nothing (Nelson 2001b, 128). Finally, a total rejection of the mechanisms creating new species is not compatible with DCM.

6. *Implications for the concept of variation.* In DCM, species are variations of the basic types, but the use of the term *variation* is inappropriate. Variation is a phenomenon within species or populations (Strickberger 2000, 657). It is misleading to use the term for the formation of new species complete with reproductive barriers. Reproductive isolation is what keeps species apart. Creationists, however, prefer to use the word *variation* to express the idea that nothing important has happened since the creation of basic types. The terms *basic types* and *ground types* are not found in textbooks. The terms operate together with variation: ground type plus variation. But all are inadequate.

7. *Implications for micro- and macroevolution.* In at least one textbook (Strickberger 2000, 648), *microevolution* is defined as changes within species. *Macroevolution* is evolution above the species level (genera, families, orders, and classes). According to DCM, considerable change—albeit ultimately bounded—may occur after the creation of basic kinds (Nelson 2001a, 684). How much change? Since genera are above the species level, DCM implies macroevolutionary processes.

8. *Implications for the rate of evolution.* The combination of a 6000-year-old earth and the number of species produced from basic types results in an astonishingly high rate of species formation: 65,000 weevil species in 6000 years amounts to more than 10 species per year.

9. *Implications for the origin of humans.* An interesting species is absent from figure 3.1. What does the model imply about the origin of humans? In traditional classification systems, humans were a separate family (Hominidae). In the modern classification based on molecular data, humans, gorillas, chimpanzees, and orangutans are placed in the family Hominidae (Futuyma 1998, 729). If creation were at the family level, traditional classification would result in the comfortable idea that humans were created separately. In the modern classification, a common ancestor of the Hominidae family would have been created, and humans, gorillas, chimpanzees, and orangutans would subsequently have evolved in a natural way.

   In both classifications, humans, gorillas, chimpanzees, and orangutans are all in the order Primates, suborder Prosimii, and superfamily Hominoidea. If a protodog could produce a family of 34 species in less then 10 million years, why should a hominoid ancestor not produce chimpanzees, bonobos, gorillas, orangutans, and humans in the same time? The chromosome variation within the hominoid group is much smaller than in Canidae (the dog family). If the genetic distance between wolf and fox were the same as that between bonobo and human, then creationists should conclude that bonobos and humans have common ancestors. Creationists, however, presume that humans are created by the direct action of a designer.

10. *Implications for the relative order of appearance in the fossil record.* In figures 3.1 and 3.2, the vertical axis is the time axis. In figure 3.1, the branches start at different times; but, remarkably, all basic types start at the same time. Where is the evidence? It contradicts the chronology of the fossil record. Furthermore, the fossil record shows that bacteria, the first eukaryotes, invertebrates, vertebrates, land plants, fishes, birds, mammals, and *Homo sapiens* did not originate at the same time in the history of the earth.

11. *Implications for the absolute times of appearance in the fossil record.* The cat family appeared 20 million years ago (Strickberger 2000, 243). The history of the horse family, including the fossil Equidae, starts in the early Eocene, approximately 55 million years ago. Neither absolute time is compatible with young-earth creationism. Moreover, the hypothetical basic types need as much evidence from the fossil record as does any other ancestor in the theory of evolution.

12. *Implications for the origin of species.* A general implication of the dynamic-creation model is that all end products—that is, all species—are not

(directly) created. Because there are no basic types alive today, no species we now encounter has been created. The beautiful ornamentation of the Argus pheasant, which Darwin (1871, 92) noted "was more like a work of art than of nature," was not created by God but by selection (see items 4 and 5 on this list). The Argus pheasant is a member of the pheasant basic type. If the common ancestor of that group did not possess the eyespots on its tail, mutation and natural selection must have created the eyespots. The stunningly ornamented birds of paradise, the tail of the peacock, the stripes of the zebra, and the human brain have evolved by mutation and natural selection.

Evaluation of the Dynamic-Creation Model

MYSTERIES

The dynamic-creation model uses standard neo-Darwinian processes when convenient but also introduces mysteries and fatal inconsistencies. Let's first have a look at the orthodoxy. The mini-trees imply common descent, branching evolution, hierarchical taxonomic levels, origin of new species, natural selection, and mutation. For example, the model explains similarities within basic types (similarities of dogs, wolves, foxes, and coyotes) by common descent.

Additionally, the differences between dogs, wolves, foxes, and coyotes are explained by divergence of the organisms arising from the ancestral basic type. Both facts are reflected in the mini-trees. So far, so good. But then a huge difference from the standard Darwinian explanation arises: in DCM, cats and dogs have an independent origin. In other words, cats and dogs are completely unrelated groups without common descent. That claim destroys the standard (Darwinian) explanation of their similarities.

DCM offers no alternative explanation, which introduces a deep mystery. I cannot stress enough how amazing it is that the model cannot answer straightforward questions such as why cats and dogs share characteristics and are placed in the same group, Carnivores, or why pheasants and ducks are placed in a group called birds. Who would deny that cats, dogs, bears, and weasels share Carnivore properties?

Darwinian theory explains their shared properties by a common Carnivore ancestor and explains their differences by divergence since the ancestral lines split. The question about similarities can be repeated for every basic type. Similarities do not stop beyond the boundaries of basic types. The whole Linnaean classification system unacceptably becomes a mystery in the dynamic-creation model.

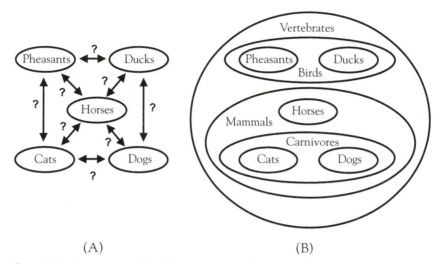

Figure 3.3. Basic types and evolutionary groups. Basic types (a) are exclusive, whereas evolutionary groups (b) are inclusive.

INCONSISTENCY

Now the logical inconsistency is easy to see. If the similarities and dissimilarities are a good reason for classifying individual organisms into the hierarchical categories, species, genera, and families, and for explaining that pattern with common descent, then why are those reasons not equally valid for higher categories such as orders, classes, phyla, and kingdoms? Why is common descent a good explanation up to the family level and a bad explanation at higher levels?

Even horses and birds share vertebrate characteristics. This pattern of similarities is called the groups-within-groups pattern, or inclusive groups (see figure 3.3b). To return to Nelson's mini-trees: the dogs are a group within the Carnivores group, the Carnivores are a group within Mammalia, the Mammalia are a group within the Craniata, and the Craniata are a group within the Animalia. The logic of inclusive groups makes it impossible to see them as independent groups. Every taxonomic group (except the highest) is included in a higher-level group. There is no such a thing as an independent group.

ANIMAL FARM

According to DCM, basic types have an independent origin. In other words, they are not connected by common descent. All basic types are equal in the sense that they are equally independent. To paraphrase George Orwell, all basic types are unequal, but some basic types are more unequal than others. Indeed, there are degrees of similarities. Although, for example, three of the basic types—pheasants, ducks, and cats—have an independent origin, creationists cannot deny that pheasants and ducks are more equal than pheasants and cats.

Why do some basic types look similar if they do not have a close evolutionary relationship? Why do some basic types look dissimilar if they do not have a more-distant evolutionary relationship? Why expect any similarities above the basic-type level at all? Any pattern of similarities of basic types is possible. Independent origin is unable to predict a groups-within-groups pattern. Higher-level groups, such as birds, carnivores, mammals, reptiles, fishes, insects, and plants, are not expected or predicted at all by a theory of independent origin. In fact, relations between basic types are in principle unknowable (see figure 3.3a), whereas evolutionary relations between groups become clearer when new information becomes available.

What we can conclude from the dynamic-creation model is that the unity of living things is a hoax and their diversity a joke. Although Scherer (1998, 206) tries to give an empirical definition of the basic type, he in fact does nothing to establish the similarities and degrees of similarity between those types. Every biologist classifies pheasants and ducks in one group, Aves (birds), and explains their similarity by saying that pheasants and ducks have a more-recent ancestor than do pheasants and cats. The explanation of the Linnaean hierarchical classification system collapses if we accept the independent origin of basic types.

PLAN OF CREATION

In Linnaeus's time, the existence of the groups-within-groups pattern was explained as the plan of creation. It was, in reality, an unintelligible fact. Without evolution, nobody could hope for a better explanation; maybe no one felt the need for it. But when a good explanation is available, it is unacceptable to fall back into the mysteries of pre-Darwinian times. Introducing mysteries and inconsistencies and destroying the explanation of all the taxonomic categories above the family level is not exactly scientific progress.

Darwin offered an elegant explanation for the groups-within-groups pattern: common descent. The creation model leaves unexplained or completely mysterious all those similarities that common descent elegantly and consistently explains. Even allowing for creation as a scientific explanation still leaves the pattern of similarities unexplained. The assumption of created ancestors does not lead to specific expectations about the pattern of life on earth such as the groups-within-groups pattern. How could it? "Darwin, after all, banished speculation about the 'unknown plan of creation' from science" (Johnson 1993, 70).

REINVENTING COMMON DESCENT

Scherer (1998) tried to express biological relationships between species by using a new systematic category: basic types. This category does not help to classify organisms belonging to different basic types. To capture the relationships between pheasants, ducks, and all the other birds, we need a basic type called birds, which is the ancestor of all the different basic types of birds. The same holds for all the mammals and all the animals. We need basic types called mammals and animals. To capture the relationships between all forms of life on earth, we need a basic type called life. And that amounts to reinventing common descent. Nelson's and Scherer's basic types are neither "basic" nor "types." Unaware of the boundaries between basic types, mutation and natural selection can go beyond their confines.

DARWIN'S INVENTION

Nelson's criticism of Ridley's illustration is odd for many reasons. The dynamic-creation model contains mini-trees. But the idea to use trees to represent the relations between species is stolen from Darwin and was certainly not invented by creationists: "Darwin, curiously, was the first author to postulate that all organisms have descended from common ancestors by a continuous process of branching. . . . A continuing multiplication of species could account for the total diversity of organic life" (Mayr 1982, 507).

Additionally, the trees in Ridley's illustration have diverging branches, while those in Nelson's illustration are vertical. This means that species are static. The dynamic-creation model is a distortion of common descent. We cannot even say that DCM is halfway toward common descent. Nelson and other intelligent-design theorists are blind to the power and purpose of common descent, which does not explain families but life itself. To limit the scope of common descent to families is like driving a plane on the road and ignoring that it is meant to fly.

New Evidence for Common Descent of Basic Types

So far I have focused on logic and explanatory power, which can be understood without detailed knowledge of genetics and biochemistry. Indeed, Darwin knew nothing about either field, but he understood the logic of evolutionary theory. In 1900, Mendelian genetics was born. Fifty years later, James Watson and Francis Crick published the structure of DNA. Another 50 years later, the complete sequence of the human genome was published. The last event signified the transformation of genetics (the study of individual genes) into genomics (the study of the genome, or the whole gene set of a species).

When scientists started comparing whole genomes of species, startling new evidence for the similarity and common descent of species began to emerge. The DNA sequences of human and mouse, for example, revealed that not only genes but also whole segments of chromosomes of mouse and human are identical. A chromosome segment of roughly 90.5 million DNA bases on human chromosome 4 is similar to mouse chromosome 5. Almost all human genes on chromosome 17 are found on mouse chromosome 11, and human chromosome 20 corresponds entirely to the bottom segment of mouse chromosome 2. A graphical and interactive representation of all the cases of synteny (correspondence of chromosome segments of two different species) of mouse and human can be found at the web site of the Sanger Institute (2002). The maps are based on Simon Gregory et al. (2002).

These special similarities mean that hundreds to thousands of genes are found in the same order in both mouse and human. This is impressive evidence for their common descent. The distribution of genes over chromosomes cannot be explained by biochemical or biological necessity. With the exception of the Y-chromosome and, to a lesser degree, the X-chromosome, no chromosome is dedicated to a special biological function such as digestion, respiration, locomotion, reproduction, or perception. The genes that control those biological functions are distributed over 20 pairs of chromosomes in the mouse and 23 pairs in the human. Consequently, the correspondences between different species cannot be explained by necessity or chance. Historical contingency is the dominant factor that produced the size, shape, composition, and number of chromosomes. Despite the many rearrangements of chromosomes since the human-mouse split, both humans and mice inherited 195 intact, conserved segments from their common ancestor.

In the creation model, human and mouse belong to different basic types. The correspondence of their chromosome segments shows, however, that mouse and human are not basic types but derived types. Similar results are found for cats, seals, cows, horses, and rats. Indeed, while I was working on the final revision of this chapter, *Science* (13 June 2003) published a special issue devoted to the tree of life. Among the many enlightening topics was the visualization of a tree of life incorporating no fewer than 3000 species (Pennisi 2003). Because of the immense amount of information packed into the tree, it is, the editor noted (not without humor), "best viewed when enlarged to a 1.5 meter diameter" (93).

Michael Behe

Not all intelligent-design advocates are like Nelson. Michael Behe (1996) claims to accept the common descent of all life: "I believe the evidence strongly supports common descent" (176). He has since repeated that statement: "I dispute the mechanism of natural selection, not common descent" (Behe 2001b, 697).

Behe's position is puzzling. He does not say why he accepts common descent. The two quotations are nearly all he has to say about it. Perhaps he does not realize the consequences of his statement. Common descent of life means that all life on earth is physically, historically, and genetically connected. It means that life is one unbroken chain of ancestors and descendants. It means that every organism inherited all its genes from the previous generation (with slight modifications). And that includes irreducibly complex systems. (See chapters 4 through 6 in this book.)

Every supernatural intervention is a violation of common descent because it means that a new irreducibly complex system was not inherited from the parents of the individual in which that system first appeared. We could not say, "I inherited all my chromosomes from my parents, except an irreducibly complex system on my X-chromosome, which has a supernatural origin." Equally, Behe cannot claim that common descent is true except when irreducibly complex systems appear. Common descent does not allow for that kind of exception because that implies a violation of the laws of genetics. Genetics is the most exact and well-established discipline in biology. Hundreds of thousands of genetic experiments have been done since the birth of classic Mendelian genetics (1900) and the birth of molecular genetics (1953). An irreducibly complex system has never suddenly appeared, whereas the kinds of mutations necessary for evolution have routinely been observed.

Phillip Johnson

Phillip Johnson is one of the leaders of the intelligent-design movement. His opinion about common descent is stated most clearly in his well-known *Darwin on Trial* (1993): "[Creationists'] doctrine has always been that God created basic kinds, or types, which subsequently diversified. The most famous example of creationist microevolution involves the descendants of Adam and Eve, who have diversified from a common ancestral pair to create all the diverse races of the human species" (68).

This is an intriguing passage for several reasons. It suggests that humans are a basic kind and were created as such. By switching from basic kind to microevolution, Johnson avoids explicitly stating that humans are a basic kind and thus created but strongly suggests that they are. If that is the case, then Johnson's idea of basic kinds is more restricted than is Nelson's dynamic-creation model. Johnson's example of basic kind is at the species level and implicitly affirms that the meaning of *microevolution* is "change within a species."

If this is Johnson's view, then by implication all species are created, and microevolution is allowed to create only minor modifications within species. Whatever the definition of basic kinds (be it on the species or family level), microevolution by definition produces no new species. Any creation model that ends up with more species than the number of species it started with needs a natural mechanism to produce new species. Creating new species is macro-evolution, according to the textbooks, but it is very difficult for a creationist to admit that he or she accepts macroevolution.

We can see that Johnson would love to believe in the special creation of humans:

> We observe directly that apples fall when dropped, but we do not observe a common ancestor for modern apes and humans. What we do observe is that apes and humans are physically and biochemically more like each other than they are like rabbits, snakes, or trees. The ape-like common ancestor is a hypothesis in a theory, which purports to explain how these greater and lesser similarities came about. The theory is plausible, especially to a philosophical materialist, but it may nonetheless be false. The true explanation for natural relationships may be something much more mysterious. (67)

An intriguing passage. Now Johnson states that the hypothesis that apes and humans share a common ancestor is plausible but may be false. He fails to make clear whether he accepts or rejects common descent and why. To state that a scientific theory may be false is nothing new: all scientific theories may be false. Certainly the idea that the true explanation for relationships may be mysterious is not a solid reason to reject common descent. With humor, Johnson remarks that "descent with modification could be a testable scientific hypothesis" (66).

But suggesting a mysterious cause for natural relationships is not "a testable scientific hypothesis." I fail to see what philosophical materialism has got to do with it. Johnson does not propose a nonmaterialist explanation. He evidently does not like the hypothesis of common descent but is unable to find good reasons to reject it and fails to present an alternative. This is science by personal preference.

William Dembski

William Dembski, the mathematician of the intelligent-design movement, published his main works after Johnson and Behe did. He is less specific, however, about common descent than are creationists like Nelson and more ambivalent about the correctness of common descent than is Michael Behe.

According to evolutionary biologist H. Allen Orr (2002), the intelligent-design movement usually admits that people, pigs, and petunias are related by common descent. But its leading theorist, Dembski (2002b, 314, 315), does not unconditionally accept common descent. He ignores Nelson's dynamic-creation model and fails to say what, for example, the similarity of apes and humans means:

> Darwinism comprises a historical claim (common descent) and a
> naturalistic mechanism (natural selection operating on random
> variations), with the latter being used to justify the former. According
> to intelligent design, the Darwinian mechanism cannot bear the
> weight of common descent. Intelligent design therefore throws
> common descent into question but at the same time leaves open as a
> very live possibility that common descent is the case, albeit for
> reasons other than the Darwinian mechanism. (315)

Dembski is right to distinguish between common descent of all life and the mechanism of evolution, but he is wrong about the relation between the two. Yes, both are part of Darwinism, but he is incorrect to suggest that natural selection and random variation are the justification for common descent. Darwin would have adopted his theory of common descent on the basis of classification alone. Common descent is inferred from data that are independent of the mechanism of evolution. Common descent itself does not imply anything about the tempo or the gradualness or the relative importance of selection in evolution.

Dembski is determined to undermine the mechanism of evolution. He hopes to debunk common descent as a logical consequence of destroying the mechanism of evolution. Significantly, 150 years after Darwin, Dembski (2002b) still has nothing better to say than a cryptic "time leaves open as a very live possibility that common descent is the case, albeit for reasons other than the Darwinian mechanism" (315). This vague remark is very similar to Johnson's (1993) mysticism: "some process altogether beyond the ken of our science" (155). Remarkably, Johnson had previously stated that "speculation is no substitute for scientific evidence" and that "Darwin, after all, banished speculation about the 'unknown plan of creation' from science" (70). It is too late for mysticism 150 years after Darwin. Biologists have something better: it is called common descent.

Both Johnson and Dembski forget that Darwin did not know about genetic mutations, he did not know Mendelian genetics, and he did not know molecular genetics. The point is that Darwin had sufficient reason to explain patterns of similarities and dissimilarities in the organic world even without knowledge of genetics. The success of Darwin's explanation did not even depend on the specifics of his theory of heredity, which turned out to be wrong. Now that we know that the genetic language of all life (how genes are translated into proteins) is not only similar but also virtually identical in all organisms, we have a magnificent confirmation of common descent.

Darwin could not have foreseen that common descent would receive such dramatic underpinnings. The specific genetic code that all living organisms use to translate genes into proteins could have been dramatically different; no chemical laws that render the current genetic code necessary have been discovered. Each created basic type could have a different genetic code without any physiological or ecological problems (Korthof 2001). Dogs and cats could have different genetic codes. Humans and apes could have different genetic codes. Yet they do not.

Common descent would be best refuted if the most closely related organisms had the most dissimilar genetic codes. (Theoretically, genetic codes can differ in gradual ways.) But all species have essentially the same genetic code. The rare and small variations of the genetic code are superimposed on common descent and follow the pattern of descent with modification. Phylogenetic trees can be constructed for those variants. The genetic code, which translates genes into proteins, is stored in DNA and subject to mutation. Variant genetic codes have very restricted effects on the organism. Nearly all possible variant genetic codes are destructive for the organism and are subject to strong selection pressure. That explains why variations are rare. Far from being an argument against common descent, as some creationists argue, they offer clues to the origin of the genetic code.

Furthermore, Darwin did not and could not construct the theory of common descent to explain the universality of the genetic code. The universality of the genetic code was discovered more than 100 years after the publication of the *Origin of Species*. Common descent thus successfully explains a completely new fact about life on earth. All those similarities will not go away, whatever Dembski's claims about the inadequacy of the mechanism of evolution.

In fact, it is extremely hard to come up with a complete and systematic alternative to common descent. Attempts to formulate a naturalistic alternative have resulted in severe problems and absurdities. For example, the Senapathy-Schwabe hypothesis of independent origin has even greater problems in explaining the properties of life (Korthof 2002, 2003).

Is There an Alternative to Common Descent?

Is there an alternative to common descent? Can there be partial common descent? The dynamic-creation model, with its created types and mini-trees, breaks the living world into arbitrary fragments, whereas common descent unifies all life. In fact, common descent unifies all disciplines of biology. The creation model does not explain the similarities between the basic types (dogs and cats), and that is a serious deficiency, because Darwin already had an elegant explanation for the similarities between taxonomic groups.

Creation restricts natural selection and mutation in an arbitrary way. Therefore, the dynamic-creation model fails to be a consistent and complete framework for dealing with biological data. It cannot replace common descent. It can be understood only as an attempt to reintroduce the Genesis kinds and not as the result of a genuine attempt to capture the diversity and unity of life. Despite the claim that it is dynamic and modern, and even though it factually contains more evolution than its formulation reveals, the model offers no progress beyond Buffon, the eighteenth-century French zoologist. It is essentially a pre-Darwinian view of life. Since no real innovative work is done by nature after the divine creation of the basic types (only variation within bounds), it is essentially a static theory.

Behe's irreducible complexity and Dembski's complex specified information likewise are inadequate to explain the similarities we see between, for example, cats and dogs. Dembski has not even made up his mind about the truth of common descent. Without an explanation of the similarities, these proponents open a real gap in their theories of life. So far nobody has produced a full alternative explanation for all the observations that common descent neatly explains.

Therefore, I can safely say that there is currently no alternative to common descent. It is the only nonarbitrary and consistent theory of descent in biology compatible with the evidence.

Chapter 4

Darwin's Transparent Box

The Biochemical Evidence for Evolution

DAVID USSERY

MICHAEL J. BEHE is perhaps best known as the author of *Darwin's Black Box* (1996). But I know Mike Behe as a biochemist and a scientific colleague; we are both interested in DNA structures. I first heard of Behe more than 20 years ago, when I was a graduate student in a biophysical chemistry group and Behe had co-authored a paper about a form of DNA called Z-DNA (Behe and Felsenfeld 1981). (The paper was famous, to some of us, because it meant Z-DNA might be more likely to occur within living cells.) In 1997, one of my students told me about a biochemistry professor who had written a book showing that complicated biochemical systems were "intelligently designed" and could not have evolved. I had no idea that the book was written by the same Michael Behe.

Both Phillip Johnson and Behe claim that none of the reviewers of *Darwin's Black Box* found fault with the science. "The reviewers say what I knew they would say: Behe's scientific description is accurate, but his thesis is unacceptable because it points to a conclusion that materialists are determined to avoid," claims Johnson (Dembski and Kushiner 2001, 38). In the same edited volume Behe agrees with this claim: "the reviewers are not rejecting design because there is scientific evidence against it, or because it violates some flaw of logic. Rather I believe they find design unacceptable because they are uncomfortable with the theological ramifications of the theory" (100).

Interested readers should look carefully through some of the reviews of Behe's book and decide for themselves whether his statement is true. (See, for example, Coyne 1996, Orr 1996–1997, Doroit 1997, Ussery 1999.) More-recent discussions on both sides of the topic appear in the April 2002 issue of *Natural History* and Pennock's (2001) edited collection. Kenneth Miller (1999) argues that Behe's irreducible complexity fails the biochemistry test. Miller, like Behe, is a Catholic. But contrary to Behe, Miller rejects intelligent design as a scientific theory because there is scientific evidence against it and also because of flawed logic; his arguments are by no means the result of his opposition to the theological implications.

Reduction of Irreducibly Complex Biochemical Systems

I have a stake in this issue: I also have written a review of Behe's book (Ussery 1999) and was critical of his science. I will explain why.

Behe (1996) defines an irreducibly complex (IC) system as "a single system composed of several well-matched, interacting parts that contribute to the basic function, wherein the removal of any one of the parts causes the system to effectively cease functioning. An irreducibly complex system cannot be produced directly" (39). As an example of IC, Behe uses the mouse-trap and claims that it needs all five components in order to function. Take away any one component, and it will not work. Before we even get into the specific examples, the mousetrap analogy itself has problems (see chapter 2). In fact, a competition has been held to develop mousetraps with fewer than the "necessary" five components, and there are many examples of mousetraps consisting of fewer parts—including some with only a single piece (Ruse 2003, 313).

Biochemistry has many IC systems, says Behe, and he uses as an example the cilium, which consists of about 250 proteins. (See Miller 1999, 140–43, for a discussion of various reduced forms of cilia.) Behe also mentions the bacterial flagellum, which is a simplified bacterial version of the cilium with about 40 proteins, as another example of an IC system. Since the flagellar system is smaller and perhaps more tractable, I will spend time here examining it in more detail.

To begin, let us consider how many parts are needed. If something is to be irreducible, then it makes sense to agree on the minimum number of components that will allow the system to function. Exactly how many parts are minimally required for the bacterial flagellum? There are two parts to this question. First, at face value, Behe says that only three IC parts are essential for function. But each of these parts is made of proteins. The second, related question, is how many proteins are essential to make a functioning flagellum. The number of proteins is much easier to quantify and verify, whereas an IC part is more difficult to nail down. So let us try to answer the more tractable second question first and then go back to the first.

How many proteins are necessary to make a flagellum? According to Behe (1996),

> The bacterial flagellum, in addition to the proteins already discussed [200 proteins for the most complicated cilia] requires about forty other proteins for function. Again the exact roles of most of these proteins are not known, but they include signals to turn the motor on and off; "brushing" proteins to allow the flagellum to penetrate through the cell membrane and cell wall; proteins to assist in the assembly of the structure; and proteins to regulate the production that make up the flagellum. (72–73)

"About forty" is vague if we are trying to figure out the minimum number of proteins. A good place to start is to look through some of the more than 100 completely sequenced bacterial genomes (anonymous 2003a) to see how many flagellar proteins are found in various genomes. Perhaps we can find a lower bound.

The common and well-studied bacterium E. coli strain K–12 has 44 flagellar proteins. Another bacterium, such as Campylobacter jejuni, has only 27 flagellar proteins. So perhaps 27 is the lower limit? But what if we find a bacterium with even fewer flagellar proteins? Can Behe's theory of irreducible complexity tell us what to expect for the lower limit? Maybe 25 proteins are necessary. On the other hand, if people report finding a bacterium that has only 23, then we need to have a careful look through the genome to see if they have missed two flagellar proteins somewhere. This makes sense: if something is irreducibly complex, then by definition *all* of the parts must be essential. If we could figure out the minimum number of proteins essential for function, the "IC number" for a given biochemical system, then it might be possible to say that Behe's idea of IC can be a useful tool for biologists looking at complete bacterial genomes. How can we determine the IC number for the bacterial flagellum? According to Behe (1996), "because the bacterial flagellum is composed of at least three parts—a paddle, a rotor, and a motor—it is irreducibly complex" (72). Evidently, only three parts are needed, but how many proteins? I have posed this question to several ID advocates and have been told that Behe is clear that you need only three parts. If each part could be made of a single protein, then evidently only three proteins are necessary. That is, an irreducibly complex flagellum can, in principle, be reduced to a mere three essential proteins.

This sounds strange. Upon closer examination, it seems that Behe is saying, on the one hand, that you cannot reduce the complex forty-protein ma-chine of a bacterial flagellum, but, on the other hand, you can perhaps find something that is still functional but has lost about 90 percent of its protein components. That is, we could go from about forty proteins to only three es-sential proteins—one each for a paddle, a motor, and a rotor. So if we look at the problem in terms of the number of proteins, the irreducibly complex ar-gument makes predictions that can easily be tested by looking at genomic se-quences and are wildly different from what is observed.

But what about Behe's argument that a bacterial flagellum, consisting now of a mere three proteins, is still irreducibly complex since it will no longer be functional if you remove any one of the components? Behe's three parts are nice descriptions, but published scientific reviews of bacterial flagella classify the flagellar protein components into *six* functional categories, not three: regu-latory proteins, proteins involved in assembly, flagellar structural components, flagellar proteins of unknown function, sensory transduction components, and chemoreceptors (Macnab 1992).

All three of Behe's IC parts fall into the third category, structural com-ponents. Thus, when Behe talks about the three IC parts of the flagellum, he is referring to only one of the six categories of flagellar proteins defined in the literature; proteins from the other five categories are not part of his IC sys-tem (that is, they are not part of the paddle, rotor, or motor). These proteins might be important for function, but you could remove them or replace them with other proteins handy in the cell, and the system would still function. Some bacteria, for example, have only half as many flagellar proteins as other bacteria.

For consistency, let us consider the three parts Behe describes as irreduc-ibly complex. Behe claims that such an IC biochemical system could not have possibly evolved, since you need all three functions (paddle, rotor, and mo-tor) simultaneously for proper function. But what if you already had each of the three components lying around, doing other functions in the cell, and then put them together? This idea can be tested by having a closer look at the com-ponents of the three different systems: are they unique to the flagellar sys-tem, or could they be used in other, non-IC biological systems?

The paddle consists of a set of proteins called flagellin, which will self-assemble; that is, if you take individual copies of the protein and mix them together, they will spontaneously polymerize to form the paddle (Yonekura et al. 2002). In practice, you need only one or a very few proteins, but the ques-tion is whether there is evidence that this protein could have evolved. I found some pertinent references in the *Journal of Molecular Evolution*, which Behe complains deals mainly with (mere) sequence comparison. (See, for example, Harris and Elder 2002.)

Sequence comparison is where the clearest evidence for evolution lies, and that is precisely where Behe does not look. I cannot overemphasize two points: first, evolution selects organisms, not complicated biochemical systems (Lewontin 2001, Mayr 2002). Second, proteins do not evolve. Rather, they are made from mRNA, which comes from the DNA sequence. Most proteins do not last very long; eventually they get chopped up and recycled to make new proteins. What is passed on from one generation to the next and what must change for evolution to happen is the DNA sequence.

I work in a bioinformatics group where every day we look at DNA, RNA, and protein sequences. I specialize in studies of bacterial genomes that have been sequenced, doing whole genome analysis. So, for me, it is very easy to have a look at the flagellin protein. As an example, one *E. coli* version of this protein consists of 595 amino acids, coded for by 1785 base pairs. How does this sequence compare to flagellin proteins in other organisms? A computer search yielded hundreds of hits, ranging from identical matches (595 out of 595 residues) for the same protein to proteins with only 193 out of 359 amino acids matching (this was from *Salmonella enterica*, which is a bacterium closely related to *E. coli*). Thus, there is a lot of variation; more than three-fourths of the sequence can be different, yet the function is still conserved.

There is good reason for this variation. The flagellar paddles stick out from the bacterium and are a prime target for the immune system if a bacterium is living inside an animal (Eaves-Pyles et al. 2001). If the flagellin sequence does not vary, the immune system, which remembers the last time it saw a flagellum, will always kill the bacterium. This is just basic natural selection. In fact, there is quite a bit of variation of flagellin, even within the same bacterium (Meinersmann and Hiett 2000). Evolution by natural selection goes on within both the bacterium and the immune system (ironically, another of Behe's irreducibly complex systems). The immune system works by generating lots of different antibodies, and then those that work are selected for, just as in Darwinian evolution (Clark 1995).

The flagellin protein is about 400 amino acids in length, and its structure can be found on the Internet, along with a link to the three-dimensional structure of the protein (Samatey et al. 2000, 2001). Interested readers can visit this web site and download the structure, rotate the molecule, and explore the available options. The function of a protein (or RNA or DNA molecule) is determined by its shape or structure. Thus, we could have two proteins with very different sequences, but if they fold into the same shape, they might have exactly the same function. It is this principle that must be understood in order to explain how we can have such large variation in sequence yet maintain the function of the protein. We can also have a very few, seemingly small changes, which have drastic effects on the function of the protein, if those changes are in the right place.

Let us now discuss the second of the three components, the rotor. It was hard to tease out the difference between the rotor components of the flagellum versus the motor part; the two are very much intertwined. There are, however, two different proteins responsible for the rotor: FliG is the rotor protein in a simple lateral flagellum, while the FliM and FliN proteins are responsible for rotors in the polar flagellum (McClain et al. 2002). A search for sequences similar to the FliG protein from *E. coli* in other bacteria found hundreds of sequences in other organisms, ranging from perfect matches to proteins containing less than a third of the amino acids in common. Once again, here is an example of large sequence variation, providing a large source of material for natural selection to choose from.

The third and final component is the flagellar motor. According to a recent review, "We know a great deal about motor structure, genetics, assembly, and function, but we do not really understand how it works. We need more crystal structures" (Berg 2002, n.p.). In my opinion, we need to better understand how this system works before we can consider evolutionary pathways.

I found a few interesting articles, however, with respect to Behe's claim of the irreducible complexity of the three components of the flagellum. For example, there may be only a loose coupling between the proton-driven motor and the rotation of the flagellum (Oosawa and Hayashi 1986). So perhaps the idea of an IC flagellum as some sort of distinct and self-contained unit is oversimplified. This reference, incidentally, was published in 1986, or 10 years before the publication of *Darwin's Black Box*.

Consider also the recent finding that we can mix and match different motors—that is, we can take a motor that is driven by sodium ions and substitute it for a functional flagellar motor that is driven by protons instead of sodium ions (Asai et al. 2003). Lots of motors in the bacterial cell do various other functions, so the flagellar motor did not have to come out of the blue at the same time as the whole flagellar complex.

Finally, what about fossil evidence of ancient flagella? Behe has claimed that an IC system somehow negates the fossil record as evidence for evolution:

> The relevant steps in biological processes occur ultimately at the
> molecular level, so a satisfactory explanation of a biological

phenomenon . . . must include a molecular explanation. . . . Anatomy is, quite simply irrelevant. So is the fossil record. It does not matter whether or not the fossil record is consistent with evolutionary theory, any more than it mattered in physics that Newton's theory was consistent with everyday experience. (Behe 1994, n.p.)

A new theory must agree with established scientific theories if it is to be widely accepted. For example, statistical mechanics predicts the macroscopic classical thermodynamics that it replaces. Similarly, Einstein's theories predict Newtonian behavior when objects are not going too fast. Surely Behe's IC system must do the same: it must be in agreement with what we observe in the fossil record, which for bacteria goes back more than 3.5 billion years (Fortey 1997, Schopf 1999, Knoll 2003). Or are we supposed to accept only present life forms as evidence?

Of course, there are lots of complicated biochemical systems in bacteria today, but the big question is whether they have always been there, as placed by an intelligent designer, or if in fact bacterial cells have slowly changed over time from simpler systems to more-complex systems. Even if we allow for bacteria to divide once a day (*E. coli* can divide every 20 minutes), there are an awful lot of replications between now and 3 billion years ago.

Behe suggests that the intelligent designer might have put all the necessary genes into the first organism. But what we see in the laboratory is that, if an organism has extra genes (genes that are not being used), they accumulate mutations fairly quickly and soon become unusable. Within a few years, the genes of bacteria would become corrupted and disappear from the bacteria's gene pool. But Behe seems to think this would not happen, even over very long periods of time. Does he think that the intelligent designer created the first cell and then sat around and waited for 3.5 billion years for humans to come along? Maybe he is right, maybe not, but this theory does not sound like science.

In summary, all three of the irreducible components of the flagellum could have evolved independently, and the flagellum could have evolved from a combination of the three independent parts rather than suddenly being created by an intelligent designer. Such coevolution is one of several alternative mechanisms for evolution of Behe's irreducibly complex biochemical systems. Similar arguments show that Behe's three other IC systems (blood clotting, the proteosome, and the immune system) consist of reducible components that could have evolved (Miller 1999, Ussery 1999, Thornhill and Ussery 2000). As a general principle, complex biochemical systems can arise from simple precursors (Ptashne and Gann 2002).

Extraordinary Claims, Anemic Evidence

Behe (1996) makes an extraordinary claim—that finding design in biochemistry "is so unambiguous and so significant that it must be ranked as one of the greatest achievements in the history of science" (232). Furthermore, the discovery of IC biochemical systems overthrows Darwinian evolution by natural selection: "It is a shock to us in the twentieth century to discover, from observations science has made [of IC systems] *that the fundamental mechanisms of life cannot be ascribed to natural selection*, and therefore were designed. But we must deal with our shock as best we can and go on. The theory of undirected evolution is dead, but the work of science continues" (Behe 1994, n.p., emphasis added).

What evidence does Behe offer to support his extraordinary claim that life is designed and that "the fundamental mechanisms of life cannot be ascribed to natural selection"? First is his argument that complex systems such as the bacterial flagellum are IC and hence cannot have evolved. But as I have shown, there are indeed plausible mechanisms that can explain the evolutionary origin of the flagellum. (See chapter 6 for additional details.)

What other evidence does Behe marshal to his defense? A supposed lack of published papers: "Even though we are told that all biology must be seen through the lens of evolution no scientist has *ever* published a model to account for the gradual evolution of this extraordinary molecular machine," writes Behe (1996, 72) about the flagellum. I did a quick search on PubMed and found 260 published articles that have the words "flagella" and "evolution" in the title or abstract. Not all of these articles describe mechanisms in a way that Behe might like, but at least some of them do, which is enough to negate his claim that "no scientist has ever published. . . . "

For example, consider a different irreducibly complex system, the immune system, and recent papers outlining its evolution. A whole new field, *evolutionary immunology*, has come to life since Behe's book was published in 1996. Out of the 4400 articles on "evolution and immunology" that can be found in PubMed, almost 2000 have been published since 1996. This hardly sounds like a dead, unprogressive field.

Finally, when Behe looked in the index of a fat biochemistry textbook, he found the word *evolution* hardly mentioned. Thus, he concluded, evolution is not necessary to understanding biochemistry. Using this same reasoning, we could claim that the atomic theory of matter was not true or at any rate not important since it is hardly mentioned in the index of biochemistry texts.

In summary, the evidence presented for rejection of the fossil record and natural selection and in favor of adopting a belief in a designer outside nature (and hence outside the realm of science) is (1) definition by fiat of an IC system, (2) an absence of articles in the scientific literature describing the evolution of biochemical systems deemed to be IC, and (3) a paucity of entries for the word *evolution* in the indexes of biochemistry textbooks. This is anemic evidence for such extraordinary claims. Yet I have talked with people who advocate intelligent design, and they simply cannot understand why their manuscripts, which contain such weak and minuscule evidence, are not published in scientific journals. They claim that the journals are obviously biased against them because of the theological implications, and publication has nothing to do with the quality (or lack thereof) of their science.

The Popularity of Intelligent Design

While doing the background work for this article, I did a Google search, typed in "Behe," and got more than 30,000 web pages. Obviously this is more material than I can handle. Why is Behe's view of biochemical evidence for an intelligent designer so popular?

Michael Ruse (2003) deals with the issue of design in nature and evolution and argues that biological organisms are clearly different from the nonliving matter around us in that they are designed. So in this sense, he is in agreement with Behe. Ruse's designer, however, is evolution through natural selection, and there is no need to invoke the supernatural (an external intelligent designer outside the system). Behe wants to return to William Paley's watchmaker analogy from nearly 200 years ago: "Behe, Dembski, and their nemesis, [Richard] Dawkins, share a desire to return to the high Victorian era, when Britain ruled the waves and science and religion could never agree" (333). Behe and Dawkins are right in their arguments that nature is designed. The question is the mechanism of design rather than whether or not things are designed. I suspect Behe is so popular in part because of his appreciation of the complexity of nature. According to Ruse,

> However one might criticize Behe's conclusions, when he speaks
> about the inner workings of the cell, his audience senses the presence
> of a man who truly loves the natural world. Say what you like in
> criticism of Dawkins, when he writes about the echolocation mechanism of the bat or about the eye and its varieties, he reveals to his
> readers an uncommon delight in the intricate workings of the organic
> world. In this Behe and Dawkins are at one with Aristotle, John Ray,
> Georges Cuvier, and of course Charles Darwin. . . . All appreciate the
> organized complexity of the natural world. (334)

Hoimar von Ditfurth (1982) says that Christians should be amazed that the miracle of evolution occurred rather than claiming that "God is what we don't know."

In 1997, I posted a web version of my review of Behe's book. Since then, more than one-half million people have visited the web page, and I have received more than a thousand E-mails about the web review over the past 5 years. About one-third of the people who write to me like my review, whereas the other two-thirds assume I must be some sort of evil atheist because I don't agree with Behe. But I am critical of Behe's IC system because it is just plain bad science. (I do not, however, think that Mike Behe's published scientific work is bad. For example, I have recently cited one of Behe's scientific papers [2000] in an article about the relative amounts of A-DNA and Z-DNA in sequenced genomes [Ussery et al. 2002]. But publishing in a peer-reviewed journal and writing a popular book are two different things.)

In my opinion, intelligent design is not good science. Since there are practically no papers published in the peer-reviewed scientific literature on this subject, I think it makes no sense to teach it as science. Indeed, to teach it as science would be dishonest.

Chapter 5

Evolutionary Paths
to Irreducible Systems

The Avian Flight Apparatus

ALAN D. GISHLICK

PROPONENTS OF INTELLIGENT DESIGN (ID) focus on biochemical and microbiological systems such as the bacterial flagellum, which, in their terms, exhibit irreducible complexity. They focus on the microscopic world, possibly because the biochemical world is advantageous to their argument. First, biochemical systems are unfamiliar to many people and thus easy to use to impress an uninitiated audience. Second, because biochemical systems structures are rarely, if ever, preserved during fossilization, they have no independent historical record through which their evolutionary history can be investigated.

As chapter 4 pointed out, ID advocates have been vague about what constitutes an irreducibly complex system. In general, it is a system that is presumed to contain three or more closely matched parts without which the system cannot function (Behe 1996). ID advocates argue that an irreducibly complex system could not have evolved—or is extremely unlikely to have evolved—by natural selection because, before an irreducibly complex system has its function, natural selection could not have favored that function (Behe 2000, Dembski 2002b). This argument is mistaken because it presupposes that functions do not shift during evolutionary history (Miller 1999; Pennock 1999, 2001). William Dembski's and Michael Behe's insistence that a system (such as the bacterial flagellum for motility) must evolve without ever shifting its function misunderstands the nature and power of evolution by natural selection.

At the core of this oversight is Dembski's (1999, 2002b) inappropriate archer analogy (see also chapter 2). Both Behe and Dembski seem to think that evolution must be shooting at a present-day, preset target from a great distance in the past. Dembski argues that the archer intelligently designs the trajectories of his arrows to hit the target from a great distance, which assumes, for example, that the remote ancestors of birds were evolving toward flight from the start.

Evolution, however, does not make such claims about the origin of complex features. Dinosaurs did not begin by evolving a route to flight, thinking, "If I give up the normal use of my hands now, my descendants will be able to fly." The archer of evolution shoots its genes only as far as the next generation. How well it succeeds depends on the morphology of the archer: any heritable feature that helps pass on those genes will survive in its offspring. In fact, evolution is much more like shooting arrows and then painting the bull's-eyes, because what is considered a hit is determined by the success of the next generation—and, contrary to Dembski, only *after* that arrow has been fired.

A system that evolved for one purpose can later be co-opted to serve some other purpose. So numerous are the examples of co-optation that biologists have coined a term for them: *exaptations* (Gould and Vrba 1982). An exaptation is a feature that originally evolved for one function but is now used for a different function. After a feature has been exapted, it may evolve, or adapt to its new function.

It is a mistake to assume that irreducibly complex systems could not have been assembled from other systems that performed functions other than their current function. It is also unreasonable to expect irreducibly complex systems to be limited to the microbiological level of organization; they should exist at the level of the organism as well.

Behe (1996, 41) has been dismissive of the organismal level, arguing that we don't know all the parts of complex organismal systems. But organismal biology and the fossil record are important, because what matters in evolution is not limited to proteins: it encompasses the ability of the whole organism to survive and produce offspring. Accordingly, if we find irreducibly complex systems at the organismal level, we should be able to investigate their structure and evolution just as well as we can in molecular systems, if not better.

Contrary to Behe, I argue that we know the parts of systems at the organismal level at least as well as we do those at the biochemical level. At the organismal level, we have the additional advantage of a historical record for those structures and functions; thus, we can observe how they were assembled for different functions and only exapted later and canalized, or developmentally and structurally locked, into a seemingly irreducible form for their current function.

Evolution often works through exaptation, assembling systems from disparate parts and not following a plan. Darwin (1859) himself realized that selective extinction removed transitional stages. Thus, when the system lacks a historical (fossil) record, it may be impossible to see how such a system evolved, and the system may actually appear unevolvable.

Paleontologists like to say that evolution erases its history. If you were to look only at the present-day world, you would have a hard time telling how different groups of animals are related because some groups, such as birds, turtles, and whales, have evolved morphologies so different from other animals that it is hard to compare them or to reconstruct the morphological path they took to get to their current state. It is like trying to figure out the origins of the government of the United States if all you have is the present government. Fortunately, we also have the historical record, starting from the republican government of Athens, to the Magna Carta, to the Declaration of Independence, to the Articles of Confederation, through the letters and writings of Thomas Jefferson, James Madison, and Alexander Hamilton, all the way to the various drafts of the Constitution and all its amendments. In understanding the origins of the U.S. government, we are not working in a vacuum. The same is true for the history of life. Evolution may have erased much of its history in living organisms, but the fossil record preserves some of the documents; with these documents, we can reconstruct the evolution of many groups of organisms. The avian flight system provides an excellent example of how an irreducibly complex system evolves by small steps through selection for different functions, a series of functional and morphological changes detailed in the fossil record of dinosaurs.

The Avian Flight System

Perhaps nothing in the biological world has captured the human imagination as much as a bird in flight. The beauty of a soaring gull, the agility of a sparrow landing in a tree, the ferocious grace of a falcon's attack all bespeak the amazing versatility of the avian flight apparatus. Scholars such as Leonardo da Vinci strove to understand and duplicate the intricacies of avian flight (Hart 1961), and some of the earliest work in zoology concerned avian flight. Perhaps what inspires us most about birds is their apparent ability to fly effortlessly. To a bird, all the intricacies of aerodynamics and wing control come naturally. Flying, however, is not as simple as it looks. In fact, we are only beginning to understand the physics and mechanics of avian flight (Rayner 1988, Norberg 1990, Goslow et al. 1990).

The key to flight is the generation of thrust through the production of a vortex wake during a downstroke, coupled with the shedding of the vortex during the downstroke-upstroke transition. The downstroke pushes the wing surface down through the air, which produces a forward thrust and generates lift. The upstroke moves the folded wing into the raised position for the next downstroke. While the upstroke is aerodynamically passive in low-speed flying, it takes on different aerodynamic functions during medium- and high-speed flight (Rayner 1988, Tobalske 2000).

For flight, a bird needs an airfoil, which is made of the feathers, and the ability to flex and extend the wing surface in order to complete an upstroke-downstroke cycle. Further, it needs to use these features in a way that makes flight possible. When coupled with an airfoil, the downstroke can generate a vortex wake that creates thrust and lift. The avian flight system is an intricate assembly of bones, ligaments, and muscles, to say nothing of the feathers. Feathers provide only the airfoil; the mechanics of the wing itself are provided by the bones, ligaments, and muscles.

To fly, an animal must be able to complete an upstroke-downstroke cycle, or flight stroke. The flight stroke includes the ability to make the transitions between upstroke-downstroke and downstroke-upstroke (or recovery stroke). To make these transitions, the animal must be able to fold and extend the wing surface. Different flying organisms have achieved this transition in different ways. Birds have a unique way of folding and extending their wings that is both coordinated and automatic. Because of the way that birds fold and extend their wing surfaces, eliminating the automatic flexion and extension system prevents birds from flying (Fisher 1957; Vazquez 1992, 1993, 1994).

This system is irreducibly complex in Behe's (1996) original sense: modern birds cannot fly unless all the parts of the automatic flexion and extension system are present and working together. (True, bats and pterosaurs fly without such a system, but the system in birds may still be irreducibly complex. The bacterial flagellum, after all, may be irreducibly complex even though humans move around without flagella.) To appreciate the point, however, it is necessary to consider the anatomy of the avian wing in detail.

Every time a bird flaps its wings, it executes a complex, interlinked series of skeletal and muscular movements. To start a flight stroke, a bird lifts its arm with its wing folded. At the top of the reach of its shoulder, it opens its wing by extending its forearm and hand (upstroke-downstroke transition). It then pulls its wing downward (downstroke), which provides the propulsive thrust of flight. At the bottom extent of the downstroke, the bird folds its wing (downstroke-upstroke transition) and lifts its arm upward (upstroke) till it reaches the upper extent of its shoulder motion. Then the cycle begins again.

The key steps in this sequence are the folding of the wing for the upstroke, which retracts the airfoil for the upstroke, and the unfolding of the wing for

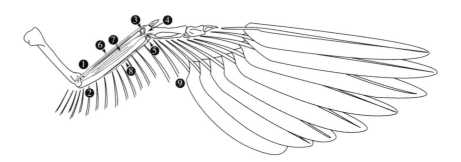

Figure 5.1. The irreducible parts of the avian flight system: (1) radius, (2) ulna, (3) radiale, (4) semilunate trochlea of the carpometacarpus, (5) ulnare, (6) M. *extensor metacarpi radialis*, (7) M. *extensor metacarpi ulnaris*, (8) M. *flexor carpi ulnaris*, (9) feathers.

the downstroke. Moreover, the wing is folded and unfolded by an automatic, coordinated flexing and extension of the wrist with the elbow. In this system, when the elbow flexes the forearm against the humerus, the wrist flexes the hand against the side of the forearm, folding the wing. When the forearm is extended at the elbow, the wrist straightens the hand, extending or opening the wing. These motions occur simultaneously, governed by a combination of skeletal morphology, ligaments, and muscles that originate on the humerus and insert onto the metacarpals. Figure 5.1 illustrates the components of the avian flight system. Automatic, coordinated flexion and extension are neces-sary for the downstroke-upstroke transition in avian-powered flight, allowing the wing to be folded up to minimize resistance during the upstroke and then re-extended for the next downstroke (Coues 1871; Fisher 1957; Dial et al. 1988; Vazquez 1992, 1993, 1994). This automatic motion is the result of a kinematic chain involving a number of key bones, ligaments, and muscles; without any of them, the bird could not fly.

Essential to this kinematic chain are the bones. The key bones are the humerus, radius, ulna, radiale, ulnare, and the semi-lunate joint surface (*tro-chlea carpalis*) of the carpometacarpus. When flexed and extended, the elbow behaves as a hinge so that the distal end of the ulna rotates inward during flexion and outward during extension. In birds, the elbow, forearm, and wrist act as a functional unit. When the elbow is flexed, the wrist is automatically flexed with it. Conversely, when the elbow extends, the wrist automatically extends. In birds, this functional unity is critical to the wing-folding mecha-nism of flight and is due to a ligamentous and skeletal link between the el-bow and wrist. The radius and ulna slide parallel to each other and drive a kinematic chain that pushes the wrist closed and pulls it open. This process has been clearly explained and well illustrated by Rick Vazquez (1992, 1993, 1994) but was originally described by Elliot Coues (1871) and later documented by Harvey Fisher (1957).

The kinematic chain starts with the enlarged radial condyle of the humerus. As the elbow flexes, the radius rides over the condyle and is pushed forward, sliding parallel to the ulna in a distal direction. This in turn pushes the radiale forward. Then in concert with a series of ligaments (*l. radiocarpo-metacarpale dorsale, l. radiocarpo-metacarpale craniale, l. radiocarpo-metacarpale ventrale*), the radiale slides along the *trochlea carpalis* of the carpometacarpus, automatically closing the wrist (Fisher 1957; Vazquez 1992, 1993, 1994). During wing extension, as the elbow opens, the radius is pulled back by the dorsal collateral ligament of the elbow, which in turn pulls the radiale back; the radiale along with the ligaments then pulls the *trochlea carpalis* back along the radiale, straightening the wrist (Fisher 1957; Vazquez 1992, 1993).

The second part of this chain involves the muscles that drive the flight stroke. The first muscle is the M. *supracoracoideus*. During the upstroke, the M. *supracoracoideus* lifts and rotates the humerus, pulling it into position at the top of the upstroke (Poore, Ashcroft, et al. 1997; Poore, Sánchez-Haiman, and Goslow, 1997). At the top of the upstroke, the triceps extends the elbow, which in turn functions with the M. *extensor metacarpi radialis* (EMR) to straighten the wrist. The EMR originates on the face and outer surfaces of the distal end of the humerus, and the muscle inserts onto the extensor process of the first metacarpal without attaching to the radius or radiale (Gadow 1888–93; Shufeldt 1898; Hudson and Lanzillotti 1955; George and Berger 1966; McKitrick 1991; Vazquez 1992, 1993, 1994). This muscle functions along with the radiale and wrist ligaments for the automatic extension of the hand with the elbow. This extends the wing for the downstroke.

The downstroke is principally driven by the M. *pectoralis* (the breast meat of the bird), the largest muscle of the arm system. Automatic flexion is governed by two muscles, the M. *extensor metacarpi ulnaris* and M. *flexor carpi ulnaris*. At the bottom of the downstroke, the biceps muscle flexes the elbow, which in turn flexes the M. *extensor metacarpi ulnaris* (EMU) and M. *flexor carpi ulnaris* (FCU) muscles. These muscles flex the hand back against the forearm, folding the wing. The EMU originates on the outside of the distal end of the humerus and inserts onto the dorsal surface of the second metacarpal of the carpometacarpus. The FCU originates on the inner side of the distal end of the humerus and attaches to the carpal bone called the ulnare (Gadow 1888–93; Shufeldt 1898; Hudson and Lanzillotti 1955; George and Berger 1966; McKitrick 1991; Vazquez 1992, 1993, 1994). The FCU acts in concert with the ulnare and the *ulnocarpo-metacarpale ventralis* ligament, which pulls the hand back against the ulna (Vazquez 1992, 1993).

The components minimally necessary to accomplish this automated upstroke-downstroke cycle have been experimentally tested (Fisher 1957; Goslow et al. 1989; Vazquez 1992, 1993, 1994; Poore, Ashcroft, et al. 1997; Poore, Sánchez-Haiman, and Goslow, 1997). The critical parts are the radiale, ulnare, radius-ulna, *trochlea carpalis*, M. *extensor metacarpi radialis*, M. *extensor metacarpi ulnaris*, and M. *flexor carpi ulnaris* (see figure 5.1). Remove any one of these skeletal or muscular components, and the system will not function.

The avian system also meets Dembski's and Behe's standard of irreducible complexity by having well-matched parts (Behe 2000, Dembski 2002b). The radius is attached by ligaments to the ulna and humerus, which guides the radius over the radial (dorsal) condyle; the radius fits against the radiale; the radiale fits into the trochlea of the carpometacarpus; and so on.

Kitchen-Counter Comparative Anatomy

The experiments used to investigate the avian flight system involve flying birds in wind tunnels, surgically altering bones, cutting ligaments, and de-innervating muscles, which prevents their function. But you can perform similar experiments yourself in your own kitchen with a sharp knife and a raw chicken wing. (The experiment works better with an entire raw bird because sometimes the wings in packages are broken.)

Any wing is divided into three segments of about the same length folded into a Z. The segment with the pointed end is the wing tip. The bone in it is the fused hand and finger bones of the bird, called the carpometacarpus. If you look closely, you can see that it has two parts of different lengths. The longer part is the fused second and third fingers. The primary feathers attach to these fingers; the hand and its feathers provide the principal lift and propulsive force of the wing (Rayner 1988, Norberg 1990). The shorter part is the first finger, or thumb. This digit carries a feather called the alula, which helps to control the aerodynamics during low-speed flying, much as do the front flaps of an airplane (Norberg 1990).

If you continue up the arm from the hand, the next segment is the forearm, composed of the radius and the ulna. The radius is the top bone on the same side of the arm as the thumb; it is thin and does not provide the principal mechanical strength of the forearm. The larger bone of the forearm is the ulna, to which the secondary wing feathers attach. The third and most robust segment is the humerus. The principal flight muscles, which raise and lower the arm as well as extend and retract the forearm, attach to this segment. (This is why it has the most meat.)

Those are the principal segments of the wing; to see how they work together with the muscles, take the wing, hold it at both ends, and stretch it out. The big triangular flap of skin stretched between the humerus and forearm is called the propatagium. It serves as the leading edge of the wing and contains the propatagial ligaments. These ligaments help to keep the leading edge taut during flight and aid in the automatic extension of the hand as described previously (Shufeldt 1898, Vazquez 1993). If you hold the arm by the middle segment and pull the humerus away from it, you will notice that the hand automatically extends as well. Push back on the humerus, and the hand flexes back against the forearm. You have just witnessed the automatic, coordinated flexion and extension of the hand with the forearm. This motion is important for the bird's wing during flight.

Now you are ready to begin experimenting. Start by cutting the propatagium and the patagial ligaments. Does the wing still automatically open and fold? If you have done it right, it will. Therefore, the patagial ligaments are not necessary for automatic flexion and extension.

To continue, next skin the wing to reveal the muscles and ligaments beneath. After removing the skin, you can start to investigate the muscles that flex the wing. The large muscle on the front of the humerus is the M. *biceps*. This muscle flexes the elbow, triggering the kinematic chain of automatic wrist flexion. The fleshy muscle that runs down the back of the humerus and attaches to the ulna's olecranon process, the bony extension of the proximal end of the ulna that is closest to the humerus, is the M. *triceps*. This muscle extends the elbow. Pull on this muscle, and watch the forearm and hand straighten; pull on the biceps, and watch them flex. These are the muscles that the bird uses to flex and extend its arm during flight.

Now look at the muscles of the forearm, the large muscle on the top of the radius and ulna. This is the M. *extensor metacarpi radialis*, which originates from the distal end of the humerus, near the biceps muscle, follows along the radius, and inserts onto the extensor process of the first metacarpal. If you cut this muscle, you will find that the hand will no longer extend with the elbow, but it will still automatically flex with the elbow. Next, examine the flexor muscles. The M. *extensor metacarpi ulnaris* is the large muscle that originates on the outer side of the distal end of the humerus, proceeds along the ulna, and inserts onto the back of the hand. The M. *flexor carpi ulnaris* is the large muscle that originates on the inner side of the distal end of the humerus and inserts onto the ulnar side of the wrist. If you cut both of these muscles, you will see that the arm can no longer automatically flex very well. Finally, if you remove the radiale, or shorten the radius, but leave the muscles, the hand will no longer flex or extend with the elbow.

So having made a mess in your kitchen and ruined a perfectly good potential hot wing, you can conclude that the avian flight system is irreducibly complex and thus—according to Behe's argument—could not possibly have evolved. Since all these components are necessary for flight, then flight was not possible before they were all assembled. Therefore, says Behe, they could not have been selected for flight.

But this assumption relies on the notion that these components evolved specifically for flight. What if they did not? What if these structures originally evolved for some other function and only later exapted for flight?

How could such a claim be investigated and tested? One major source of data is the fossil record. By looking at the history of avian flight, we can see how the flight apparatus was assembled, not for flight but for predation and insulation. Limbs and feathers were employed together to function in a rudimentary form of flight, later modified into the highly refined form that we see today.

Fossils, Feathers, and Flight

We can use comparative anatomy to investigate the morphology of ancestors even when the ancestors are not preserved in the fossil record. Paleontologists employ a technique called phylogenetic bracketing (Witmer 1995, Bryant and Russell 1992). Despite its fancy name, the idea of phylogenetic bracketing is relatively simple. At its core, it presumes that any features shared by two or more related animals were also present in their last common ancestor. This hypothesis may be rejected if evidence to the contrary is discovered during the investigation of a specific case. With a phylogenetic bracket and some fossils, we can determine when certain features arose along the line to avian flight.

Using evidence from the fossil record, we now know that birds are living dinosaurs that descended from cursorial (running), bipedal predators (Ostrom 1974, 1976, 1979, 1997; Gauthier and Padian 1985; Gauthier 1986). The fossil record of the evolution of avian flight is extensive and constantly growing; in particular, we now have a detailed record showing how the skeletal and muscular systems were modified along the route to flight. What we see in this record is that all of the skeletal, ligamentous, and muscular features just discussed arose gradually along the lineage leading to birds. We can further infer that those features arose not for flight but in conjunction with predation (Gauthier and Padian 1985; Gishlick 2001b, 2001c).

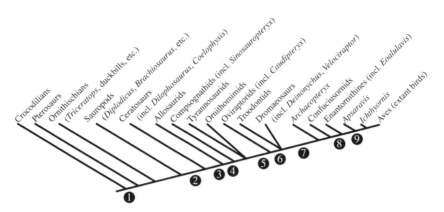

Figure 5.2. A simplified phylogeny for birds, plotting the appearance of characters relevant to flight: (1) four-chambered heart; (2) four-fingered hand; (3) three-fingered hand, showing fused distal carpals 1 and 2 with a trochlea restricting the motion in a lateral plane; (4) filamentous feathers, longer hands; (5) enlarged semilunate carpal, with M. *extensor metacarpi ulnaris* inserting onto back of hand (automatic flexion), pennaceous feathers on hands and tail; (6) M. *extensor metacarpi radialis* insertion on flange of metacarpal 1 (automatic extension); (7) flight feathers; (8) alula; (9) fully integrated and irreducibly complex flight system.

To see how these features were assembled, paleontologists explore the origin of flight-related features in the context of the phylogeny. The phylogeny, or evolutionary relationships, of the organisms under study provides a framework on which to trace the order and evolutionary timing of the appearance of flight-related features (Padian 1982, 2001; Padian and Chiappe 1998b). Figure 5.2 shows a simplified phylogeny of birds and dinosaurs with the changes in the "irreducible" components of the flight system plotted when they occur. A myriad of detailed anatomical changes in the forelimb occur along this lineage; we will deal with only a few significant features here.

The theropod ancestors of birds were bipedal predators; they did not use their arms for locomotion but for seizing prey. This conclusion is based on numerous independent phylogenetic tests based on comparative anatomy (Ostrom 1976, 1979; Gauthier and Padian 1985; Gauthier 1986). This phylogenetic context gives us an independent way to test the adaptive route to avian flight (Padian 1982, 2001; Padian and Chiappe 1998b). Any explanation for how the avian flight system arose requires an explanation of how theropod arms were used in a way that enabled them to acquire features that could then be exapted for flight. Basal theropods, those lower on the tree (see figure 5.2), such as *Dilophosaurus* or *Coelophysis*, had four-fingered hands and largely moved their hands in palmar-antipalmar direction (the direction we move our hands at the wrist when dribbling a basketball), as opposed to birds, who have three-fingered hands and move their wrists in a radial-ulnar direction (try moving your hands in the direction of the thumb or pinky). By comparison, the principal direction of manual movement in humans is in the palmar-antipalmar direction, with a high degree of pronation and supination: you can turn your hands more than 180 degrees, either palms up or palms down.

Birds and other theropods cannot pronate or supinate much, and they do not do so by crossing the radius and ulna as in humans. They hold their hand so that the plane of the hand is perpendicular to the ground. To move your hand like a bird's, hold your arms at your sides, bent at the elbows, with your hands out in front of you. Now turn your hands so that your palms face each other with your thumb pointing up and your pinky pointing down. If you were a bird, you could bend your hand at the wrist only in the direction of your thumb or the direction of your pinky, and you could not rotate your hands so that your palms can be up or down. You don't have a lot of range of motion in that axis (radial-ulnar), but birds and the theropods closest to them can move their wrists only along that axis.

The first flight-related feature that evolved on the line to birds is the *trochlea carpalis* of the wrist, which the radiale slides along. The evolution of the trochlea leads to the restriction of the wrist motion into a lateral (radial-ulnar) plane. The appearance of the trochlea along with the initial wrist motion restriction occurs at the base of the group called tetanurines (figure 5.2) and is accompanied by a reduction to three fingers, as seen in avians. This is first apparent in torvosaurs (Gauthier 1986); the *trochlea carpalis* becomes deeper in allosaurids (Gilmore 1915, Ostrom 1969, Madsen 1976, Chure 2001). A good example of this simple trochlea can be seen in theropods such as *Allosaurus*, where there is an enlarged first distal carpal fused to a small second distal carpal. Together these bear a shallow trochlea along which a somewhat enlarged radiale can slide (Madsen 1976, Gauthier 1986, Chure 2001). These bones enabled a degree of radial-ulnar transverse motion between the distal carpals and radiale. The distal carpal trochlea was not very deep, however, so it still allowed the ancestral dorsal-palmar motion between the carpals. As you move up the tree (figure 5.2) toward birds, the distal carpals continue to enlarge, increasing the range of lateral motion in the wrist, along with a steady enlargement of the radiale and deepening of the trochlea. The arms are steadily lengthened, particularly the hand.

The next important feature occurs within a group called the Coelurosauria. It is here that the first feathers evolve. When feathers first appear, they are simple and hairlike, much like the natal down of chicks (Sumida and Brochu

2000; Prum and Brush 2002, 2003). These are first visible on the small compsognathid dinosaur *Sinosauropteryx* (Chen et al. 1998). The feathers most likely served as insulation, much as they do in juvenile birds today (Brush 2000; Prum and Brush 2002, 2003), rather than being used for flight. Phylogenetic bracketing indicates that all dinosaurs phylogenetically above *Sinosauropteryx* (that is, to the right of *Sinosauropteryx* in figure 5.2) would have inherited at least these simple feathers from a common ancestor. Imagine a fuzzy *Tyranno-saurus rex*, at least a juvenile one. (This was nicely illustrated in the November 1999 issue of *National Geographic*.) In basal coelurosaurs, the hand is longer, approaching the long hands of maniraptors and avians (Ostrom 1974, Gauthier 1986).

Many of the important changes in the forelimb occur within the mani-raptors. In maniraptors, we first see a large semi-lunate surface of the fused distal carpals, as in avians, along with a deep trochlea and large triangular radiale. Further, we first see an avianlike range of motion for the wrist and forelimb. Such a range of motion is first seen clearly in theropods such as *Oviraptor*, but it was probably present in the more basal theropod *Ornitholestes* as well.

At this point in the phylogeny, we also see evidence of an EMU insertion on the dorsal surface of the second metacarpal, indicative of the automatic flexion system seen in birds (Gishlick 2001a). Thus, the automatic flexion of the forelimb may have appeared before automatic extension. Pennaceous feathers (which have a rachis and vanes, what we think of as true feathers) appear at this level in the phylogeny, as exhibited on the hands and tails of the oviraptorosaurs *Caudipteryx* and *Protarchaeopteryx* (Ji et al. 1998; for good pictures, see the July 1998 issue of *National Geographic*). All theropods following this point in the phylogeny, therefore, would have had true feathers on their hands and tails. These feathers are symmetrical, indicating that they were not used for flight: flight feathers are asymmetrical for aerodynamic efficiency.

True feathers evolved for some function other than flight, perhaps as a display feature or for brooding behaviors (Padian and Chiappe 1998a, 1998b; Prum and Brush 2002, 2003). Or perhaps the first feathers were employed as "blinders," causing the hands and arms to appear larger and thus prevent small prey from dodging to either side, or employed to herd prey (Gishlick 2001b). Feathers on the hands would have also had aerodynamic properties and may have aided in thrust during running (Burgers and Chiappe 1999; Burgers and Padian 2001) or traction while running up inclined surfaces (Dial 2003). Further, because of the way the feathers attach to the hand, they would not have greatly inhibited the grasping ability of the forelimbs (Gishlick 2001b, 2001c).

If you continue further into the Maniraptora, troodontids and dromaeosaurs (such as *Velociraptor* and *Deinonychus*) evolve an enlarged flange on the first metacarpal for the insertion of the EMR, giving the first evidence for the automatic extension of the hand with the elbow in theropods (Gishlick 2001a). The very constrained motion of the forelimb at this point in the phylogeny could be regarded as a rudimentary form of the flight stroke previously detailed. Thus, the motions of avian flight and its muscular control evolved before flight and did not appear all at once, perhaps under selective pressure for grasping of prey (Gauthier and Padian 1985, 1989; Padian and Chiappe 1998a, 1998b; Gishlick 2001b, 2001c).

Archaeopteryx displays the first evidence of powered flight in dinosaurs, indicated by its asymmetrical feathers (Feduccia and Tordoff 1979, Rietschel 1985), large wing surface capable of supporting body mass (Padian and Chiappe 1998a, 1998b), and the capability of elevating the shoulder for the upstroke (Jenkins 1993). But even at this point, *Archaeopteryx* still maintained a functional grasping hand (Gishlick 2001b, 2001c). From this point on, however, selection for flight becomes stronger than selection for a grasping hand, and thus the flight system starts to become fine-tuned. By the time of *Confuciusornis*, we see the beginning of the fusion of the wrist elements into a carpometacarpus as well as the loss of a functional (for grasping) second digit (Chiappe et al. 1999). As the functional demands of flight increased, however, it became impossible to maintain dual use (Gishlick 2001b, 2001c). This transfer of function from grasping, through a grasping-flying stage, to pure flying is documented by the sequence of structures found in the fossils of *Deinonychus*, *Archaeopteryx*, *Confuciusornis*, and *Eoalulavis*, in that order. With the addition of an alula (Sanz et al. 1996)—one grasping finger is not particularly useful—the hand became used solely for flight, and the third digit was reduced and fused to the second thereafter. The alula modifies the flow of air over the wing to reduce turbulence. It allows birds to improve low-speed flying and gives them the maneuverability that make them the lords of the air. From here on, the avian flight system is something that can be considered irreducibly complex. Hence, we may conclude that irreducible complexity does not imply unevolvability.

By no means did the avian flight mechanism assemble all at once in its irreducible form. Rather, it was assembled piecemeal over millions of years and millions of generations. A grasping strike happened to produce thrust when the wing was large enough. The kinematics worked in the right way, so the feathered arm was exapted for flight rather than selected for flight originally. The historical record has enabled scientists to dissect such irreducible structures into reducible components and allows us to understand how supposedly irreducibly complex structures can evolve. With that in mind, we ask, what is the biological significance of irreducible complexity?

The answer, I think, is not much. Irreducible complexity would be biologically significant if the proponents of intelligent design were correct that it meant "unevolvable." But the example of the avian flight system contradicts their central claim. Irreducibly complex systems are evolvable, and that evolvability can easily be documented when a fossil record is available to put the structure into a historical context. I would bet that if a fossil record were available for the bacterial flagellum, we would see the same type of exaptation and mosaic evolution that we see in the avian-flight system. It's that simple.

When Behe (1996) proposed irreducible complexity, he treated it like a fundamental property of the biological world that evolutionary biologists had been intentionally or unintentionally not acknowledging since Darwin. That was not the case. Irreducibly complex systems have not been missed; they just were not considered insurmountable obstacles for evolution because of exaptation and mosaic evolution. The example of the avian flight system shows clearly that it is not evolutionary biologists who are missing something fundamental.

Chapter 6

Evolution of
the Bacterial Flagellum

IAN MUSGRAVE

THE BACTERIAL FLAGELLUM is an organelle that looks strikingly similar to a machine constructed by humans (Namba et al. 2003). This similarity has led to claims that it is a construct rather than a product of evolution. Indeed, the bacterial flagellum has become the mascot of the intelligent-design movement: it is one of only two examples of alleged design considered in any depth, graces the cover of William Dembski's *No Free Lunch* (2002b), and features in a recent video promoting intelligent design. Yet "the" bacterial flagellum does not exist.

The image that graces Dembski's book is a representative flagellum of eubacteria, one of the two fundamental subdivisions of prokaryotes (bacteria in general). Archaebacteria, the other fundamental prokaryote group, have flagella that are superficially similar but do not look like a human-constructed machine. Given the importance that "the" eubacterial flagellum has assumed in debates over intelligent design, it is worthwhile looking at "the" flagellum in more detail.

In this chapter, I outline the construction and function of eubacterial and archaebacterial flagella and their relationship to other systems. I discuss some of Dembski's objections to current accounts of flagellar evolution and end with a possible scenario for the evolution of eubacterial flagellum, based on structures and functions in known, related bacteria.

Michael Behe (1996) has listed the eubacterial flagellum as one of the systems that he believes is irreducibly complex and unable (or unlikely) to have been produced by evolution. Building on Behe's claims, Dembski (2002b, 289) has made the eubacterial flagellum a central point of his key chapter, "The Emergence of Irreducibly Complex Systems," and has produced an analysis of the flagellum by assuming that all elements of the flagellum arose randomly. He claims that his analysis supports the intelligent design of the flagellum.

Kenneth Miller (2003) and David Ussery (chapter 4 in this book) have addressed key aspects of our understanding of the flagellum and its evolution. Dembski (2003) has not found such accounts convincing. First, he does not seem to understand that the eubacterial flagellum is only one of a range of motility systems in bacteria—systems, moreover, that revolve around a common thread—and that motility is just one function of the flagellum. Further, he artificially categorizes the eubacterial flagellum as a machine. By viewing the eubacterial flagellum as an isolated outboard motor rather than a multifunctional organelle with no explicit, human-constructed analog, Dembski makes the problem of flagellar evolution artificially and misleadingly difficult.

Is the Flagellum Evolvable?

Dembski's calculation method for deciding whether or not systems are designed first requires elimination of systems assembled by natural laws such as natural selection. Given our finite state of knowledge, there is always the possibility that if we currently do not have an explanation due to natural laws, we may find one in the future (Wilkins and Elsberry 2001). To avoid this problem, Dembski (2002b) attempts to provide a proscriptive generalization that will eliminate any explanation based on natural law and then also allow him to eliminate chance hypotheses. He gives as an example of a proscriptive generalization the second law of thermodynamics, which proscribes the possibility of a perpetual motion machine (274).

To provide a proscriptive generalization with regard to the flagellum, Dembski accepts Behe's (1996) description of the flagellum as irreducibly complex (IC). He claims that the flagellum, considered as a system of motor, shaft, and propeller, cannot be built sequentially. He thus claims that describing the flagellum as IC eliminates natural selection as a possible mechanism and proceeds to his calculations to eliminate chance. There are, however, two problems with using the alleged IC nature of the flagellum that are not covered in his book.

First, while allegedly eliminating directly evolved systems (see, however, chapters 4 and 5 in this book), Behe (1996) himself points out that these systems may evolve indirectly: "Even if a system is irreducibly complex (and thus cannot have been produced directly) . . . one can not definitely rule out the

possibility of an indirect, circuitous route. As the complexity of an interacting system increases, though, the likelihood of such an indirect route drops
precipitously" (40). IC by itself does not provide the proscriptive generalization that Dembski requires.

Second, the specification of an outboard motor, which provided the IC
system description of motor, shaft, and propeller, is a flawed human analogy
for the actual flagellar system. Thinking in terms of human design has misled
Dembski. Indeed, in terms of Dembski's (2002b) modification of Behe's original definition, "A system performing a given basic function is irreducibly complex if it includes a set of well-matched, mutually interacting, non-arbitrarily
individuated parts such that each part in the set is indispensable to maintaining
the system's basic, and therefore *original*, function. The set of these indispensable parts is known as the irreducible core of the system" (285, emphasis
added).

The flagellum is probably not IC at all because the original function of
the eubacterial flagellum, which can survive massive pruning of its components, is almost certainly secretion, not motility (Hueck 1998, Berry and
Armitage 1999, Aizawa 2001). To explain this claim, I will examine the variety of motility systems found in bacteria and show that the eubacterial flagellum functions is far more than just a motility system.

Motility Systems in Prokaryotes

Discussions of the eubacterial flagellum often give the impression that it is
the only motility mechanism in bacteria. Not all prokaryotes move, however,
and not all motile prokaryotes use flagella. Furthermore, the motility methods in the two domains of prokaryotes—eubacteria and archaebacteria—are
very different, even though they look superficially similar. There is a range of
motility systems even within the eubacteria themselves. Entire groups have
no flagella but still manage effective swimming; others use gliding motility
across surfaces. I will briefly summarize the basic mechanisms used in gliding
and swimming motility; as we will see, they throw light on the origin of the
flagellum.

Table 6.1 shows different motility systems in prokaryotes. It is by no means
exhaustive, and there are probably undiscovered motility systems. What is clear
from the table, however, is the extensive role of secretion in motility. This
role is greater than it appears at first glance because many systems that look
unrelated to secretion are in fact rooted in that function.

In gliding motility, almost all the mechanisms are related to secretion:
the bacteria glide along a trail of secreted material in a manner reminiscent

Table 6.1
Mechanisms for motility in prokaryotes.

Gliding Motility	
Cyanobacteria	Generally carbohydrate secretion (in many species long filaments guide the secreted carbohydrates)
Myxococcus xantus	Type A motility, slime secretion
Many eubacteria	Type IV pili, ATP-driven long filaments that pull bacteria along by contraction
Flavobacterium (previously *Cytophaga*)	Slime-secreting rotatory motors

Swimming Motility	
Eubacteria	Proton-motive force-driven rotation of helical flagellar filaments
Archaebacteria	ATP-driven rotation of helical flagellar filaments (not related to eubacterial flagella)
Synechococcus (cyanobacteria)	Ca^{2+}-dependent beating of straight filaments (related to guide filaments in gliding cyanobacteria)

of slugs. Further, the slime-secretion systems in gliding cyanobacteria bear a strong resemblance to type-III secretory systems (Spormann 1999); in many gliding eubacteria, the secretory systems rotate and are driven by proton-motive force, as are the eubacterial flagella (Pate and Chang 1979).

The apparent exception is the motility produced by the type-IV pilus (see figure 6.1), where a long, whiplike filament is used to pull the bacterium along as the pilus attaches to a surface, contracts, then releases and extends to attach to a surface again. (This mechanism is also called twitching motility.) This exception is only apparent because the type-IV pilus is related to the type-II secretory systems (Thomas et al. 2001). They use the same motor systems, and the pilus seems to be an elaboration of the protein-transport apparatus. The type-II system (figure 6.1) uses an extension-retraction system to export proteins across the membrane.

Let us now look at two of the swimming systems: cyanobacterial nonflagellar swimming and the archaebacterial flagellum. Cyanobacterial swimming uses a very simple system consisting of one, but no more than two, components. Cyanobacterial swimming is due to coordinated movement of a semi-rigid, calcium-binding filament in the outer surface of the cyanobacterial coat. Interestingly, the semi-rigid filaments that function as oars appear to be modifications of the protein that guides slime from the slime nozzles of gliding cyanobacteria (Samuel et al. 2001).

Archaebacterial flagella are instructive because they look superficially

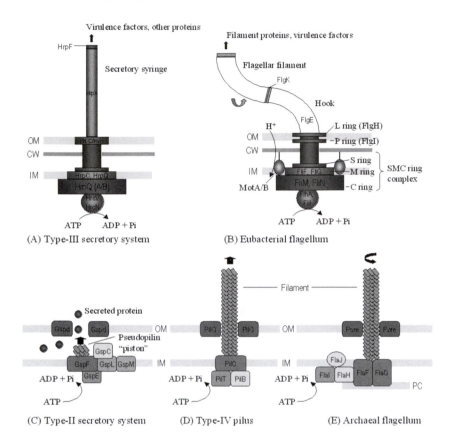

Figure 6.1. Secretory systems compared with eubacterial and archaebacterial flagella. There is significant sequence homology and functional similarity between the type-III secretory system and the eubacterial flagellum (a, b), and between the type-II secretory system, the type-IV pilus, and the archaebacterial flagellum (c, d, e).

a. The type-III secretory system is a hollow "rivet" that passes through the bacterial inner membrane (IM), the cell wall (CW), and the outer membrane (OM). It has a ring complex anchoring it to the inner membrane, which also houses the ATP-driven secretory apparatus. Another ring provides a bushing though the outer membrane. In some type-III systems, such as the Hrp system of the plant pathogen *Pseudomonas syringia* (illustrated) and some species of *Yersinia*, *Escherichia*, and *Shigella*, a hollow pilus is attached to the outer ring.

b. The eubacterial flagellum has essentially the same structure as the type-III secretory system, but with the motor protein MotAB (related to the Tol-Pal, Exb-TonB secretory motors) added. Proton transport by MotAB turns the flagellum by interacting with the SMC ring and is also responsible for some of the secretion through the hollow flagellum. There is extensive homology between the proteins of the type-III secretory system. In the rings, for example, FliF, FliG, FliN, and FliM are homologous with HrpC, HrpQ, and HrcQ(A/B); in the hook-pilus, FlgK and FlgE, with HrpF and HrpX; in similar to eubacterial flagella. They are constructed in an entirely different manner, however, and are significantly simpler (Thomas et al. 2001; see figure 6.1). Unlike the eubacterial flagellum, which can be described as motor, shaft, and propeller, the archaebacterial flagellum consists of a motor and combined shaft-propeller. This shows that the alleged three-part IC system of the flagellum can indeed be simplified. Currently only 8 to 10 archaebacterial flagellar proteins are known, although it is likely that more remain to be discovered.

There is no homology between the flagellar proteins of the eubacteria and archaebacteria (Thomas et al. 2001). The archaebacterial flagellum is, however, homologous to the type-IV pilus, which is responsible for twitching motility (see figure 6.1). They use similar motor proteins, assembly proteins, and chemical-sensing pathways (Thomas et al. 2001). Thus, there is a path in the development of the archaebacterial flagellum: from a secretory system, to an organelle for rotatory swimming motility, through a functional intermediate.

There is a clear link between secretory systems and motility, from simple gliding systems to more-complex swimming systems. This link is important in understanding the eubacterial flagellum because it, too, is a secretory system.

Structure of the Eubacterial Flagellum

Just as "the" bacterial flagellum does not exist, there is no "the" eubacterial flagellum either. Within the eubacteria there are at least two, possibly three, flagellar systems (Asai et al. 1999, Berry and Armitage 1999), based on whether their motor systems run on protons or sodium and on the complexity of the

---

Figure 6.1. (Continued)
the secretory system, FlhA and FliI, with HrcV and HrcN; and in secretion control proteins (not shown) such as the sigma factors and hook/pilus-length control proteins (Aizawa 2001, He 1997).

 c. In the type-II secretion system, ATP hydrolysis by GspE drives a piston motion of the pesudopilins anchored to the GSpF protein, which drives secretory proteins through a pore made by GspF.

 d. In the type-IV pilus, ATP hydrolysis by PilT drives contraction and extension of a whip formed of pilins anchored to PilC. The pilins exit the outer membrane by a pore formed by PilQ.

 e. In the archaebacterial flagellum, ATP hydrolysis, presumably by FlaI, drives rotation of a whip formed of flagellins anchored to FlaF and FlaG, which in turn are anchored to the polar cap (PC). The pseudopilins, pilins, and flagellins share homologous N-terminals, and GspE, PilT, and FlaI are homologous. Unlike the eubacterial flagellum, the archaebacterial flagellar filament is not hollow. For clarity the homologous prepilin peptidases, which help assemble the plunger, pilus, and flagellum are not shown (Thomas et al. 2001, Nunn 1999).

flagellar whip. Within these groupings, the structure of the flagellar elements varies; however, there is a common structure (see figure 6.1). The eubacterial flagellum has a helical filament (propeller), a hook (universal joint), a rod (drive shaft), an S-P ring (bushing around the rod, but only in Gram-negative bacteria), and the SMC-ring complex, which is the motor, includes the stator and the rotor. The entire assembly is hollow, including the actual filament. The significance of this fact will become apparent later in my discussion.

The rotor, hook, and filament are made of (nonidentical) helical proteins that self-assemble to form hollow cylindrical structures. The filament cylinder is helical, so it acts as a screw propeller when it rotates. Many eubacteria can switch the direction of rotation of the propeller (and hence the direction of travel), and the switch mechanism appears to be part of the motor complex. Eubacteria also have a chemical-sensing system that regulates the activity of the flagellum so that they swim toward or away from a chemical stimulus.

Between 30 and 50 genes are involved in the construction and regulation of the canonical eubacterial flagellum (44 in the case of the *Salmonella typhimurium* and *Escherichia coli* flagella, but only 27 in the case of *Campylobacter jejuni*); only 18 to 20 form the actual motor-switch-shaft-propeller complex.

Homologies with the Type-III Secretory System and Other Systems

As we have seen, there is a deep link between secretory systems and bacterial motility, and there is a strong link between the eubacterial flagellum and the type-III secretory system. Bacteria have multiple secretory systems, and we have already encountered the type-II system. Type-III secretory systems are involved primarily in secreting proteins that allow bacteria to attack and invade eukaryotic cells.

The type-III secretory system forms a rivet structure identical to the rod and SMC-ring complex of the flagellum (Hueck 1998, Berry and Armitage 1999, Macnab 1999; see figure 6.1). Proteins exported by this system are shunted through the hollow SMC ring and through the rod to the outside of the cell (Hueck 1998, Berry and Armitage 1999, Macnab 1999). In flagellum assembly, flagellins and hook proteins are shunted to the outside of the cell via the rod-and-ring complex. The proteins attach to the outer rim of the rod and self-assemble into a tubular structure that will become the hook and filament, and flagellar proteins pass through this tube as it grows (Hueck 1998, Macnab 1999). Thus, the flagellum and the type-III secretory system share the same structure and function. This is no mere resemblance. Homology studies show that many of the flagellar proteins are related to parts of the type-III protein-secretion system (Hueck 1998, Berry and Armitage 1999, Macnab 1999), and the majority of the homology is in the rivet structure and the secretory apparatus (see figure 6.1).

Furthermore, the genes for the rivet rod and the SMC-ring complex form a single transcription unit in both the type-III secretory systems and the flagellum. The orientation and order of these genes in the transcription unit are very similar between the type-III secretory systems and the flagellum (Hueck 1998, Berry and Armitage 1999, Macnab 1999). Finally, phylogenetic studies suggest that type-III systems share a common ancestor (Aizawa 2001).

Importantly, the switching-torque generation system of the flagellum has homologs in virtually every type-III secretory system examined so far (Hueck 1998, Berry and Armitage 1999, Macnab 1999). Intriguingly, several type-III secretory systems have tubular structures attached to the rod. *E. coli* has a filamentous structure attached to one of its type-III secretory systems, which has significant similarity to the flagellar filament (Sekiya et al. 2001). While secretion in the flagellum is closely linked to flagellar assembly, it also plays a wider role. For example, pathogenic *E. coli* use the flagellar system to secrete enzymes that attack cell walls (Young et al. 1999).

While there is no apparent homolog of the motor (MotAB) in type-III secretory systems, the motor is homologous to the motor of the Tol-Pal and Exb-TonB secretory systems (Cascales et al. 2001). This homology links MotAB and the flagellum to a wide range of secretory systems, including the carbohydrate-secretory systems used in gliding motility (Youderian et al. 2003). Major components of the gliding-secretory systems of *Myxococcus xanthus* are also related to the motor components of the Tol-Pal secretory system (Youderian et al. 2003). Like the secretory and gliding-secretory systems (where deletion of the motor proteins stops secretion), deletion of MotAB not only paralyzes the flagellum but also significantly reduces secretion through the flagellum, emphasizing the dual role of the system (Young et al. 1999).

While the type-III secretory system does not have a chemical-sensing system like the eubacterial flagellum, close homologs of this system are present in the type-IV twitching-motility system and gliding-motility systems (Spormann 1999, Thomas et al. 2001).

Thus, there are deep links between the structure of the eubacterial flagellum and secretory systems. Flagella share the same basic structure as secretory systems, they secrete proteins as do secretory systems, motors that power secretory systems power them, and they are regulated by chemical-sensing systems that regulate other secretory systems. Indeed, Macnab (1999) considers flagella to be specialized type-III secretory systems.

The Eubacterial Flagellum in Context

Looking at the context of the bacterial flagellum gives us further insight into how the flagellum arose. Flagella are often thought of exclusively as swimming-motility organelles, yet they have a wide range of other functions. First and foremost is secretion. As we have seen, the flagellum secretes the subunits that form the hook and filament parts of the flagellum. But the flagellum also secretes nonflagellar proteins of importance to bacteria (Young et al. 1999).

The next function is adhesion. The flagellum attaches bacteria to surfaces; this is important for forming biofilms (Watnick et al. 2001), which allow cells to exploit resources on surfaces. Indeed, the ability of the flagellum to bind to cells is critical for pathogenic bacteria to attach to their host cells to attack them (Giron et al. 2002). Even nonmotile pathogenic bacteria express flagella that are crippled in terms of swimming (Andrade et al. 2002), presumably due to the role of flagella in adhesion and invasion of host cells. Importantly, flagella are central in organizing bacteria into a mass to produce a nonswimming form of motility called swarming (Kirov et al. 2002).

Dembski has said that the specification for the eubacterial flagellum is an outboard motor, but as we can see, the flagellum is, at the same time, a bilge pump and an anchor (to continue the nautical theme). If we view this organelle simply as an outboard motor, we have a distorted view of what it is and what it does.

When viewed as a swimming structure, the flagellum is IC. Remove the motor, and it stops functioning; remove the hook (universal joint), and it stops functioning; remove the filament, and it stops functioning (although in some bacteria removal of the filament results in weak motility). Viewing the flagellum as an outboard motor—and an IC motor at that—provides no insights into the origin or functioning of this structure.

But view it as a secretory structure, and it is not IC. Remove the filament, and it still works; remove the hook, and it still works; remove the motor, and it still works—not as well as with the motor, but it still works. But which, in Dembski's terms, is the *original* function? Secretion plays a crucial role in this organelle, and you can't make flagella without secretion, so secretion must be the original function.

This conclusion is backed up by the crucial role that secretion plays in other motility systems. Indeed, secretion is a common thread in all motility systems described so far. This is because one of the fundamental problems of a swimming system is how to get the structures that will be used as oars or propellers through the cell wall. Secretory systems, which are fundamental to the functioning of bacteria, have already solved this problem and would be needed to get the swimming structures across the cell wall. Therefore, it is understandable that evolution would build motility systems on top of existing secretory systems. Thus, Dembski's analysis is deeply flawed.

Dembski and Type-III Secretory Systems

When Dembski (2002b) wrote section 5.10, "Doing the Calculation," he was unaware of the proposal that eubacterial flagella were related to type-III secretory systems. Miller (2003) has written a critique of Dembski's chapter based on the flagellum's relation to the type-III secretory system. Dembski (2003) has written a response to this critique and dismissed the link between type-III secretory systems and the flagellum. Among other things, he says,

> Miller doesn't like my number 10^{-1170}, which is one improbability that I calculate for the flagellum. Fine. But in pointing out that a third of the proteins in the flagellum are closely related to components of the TTSS [type-III secretory system], Miller tacitly admits that two-thirds of the proteins in the flagellum are unique. In fact they are (indeed, if they weren't, Miller would be sure to point us to where the homologues could be found). (n.p.)

In fact, they are not. While Miller emphasized the type-III secretory system, we now know that between 80 and 88 percent of the eubacterial flagellar proteins have homologs with other systems, including the sigma factors and the flagellins (Aizawa 2001; also see chapter 4 of this book). Homologies between a few of the rod proteins and nonflagellar proteins have not been found yet, but they appear to be copies of each other and related to the hook protein. In the end, there is not much unique left in the flagellum. As I have pointed out, the motor proteins and the chemical-sensing system have homologs in other secretory systems. This means that the very functions that Behe and Dembski think are IC have common ancestry with similar functions in other bacteria.

Dembski (2003) also writes,

> But let's suppose we found several molecular systems like the TTSS that jointly took into account all the flagellar proteins (assume for simplicity no shared or extraneous proteins). Those proteins would be similar but, in all likelihood, not identical to the flagellar proteins (strict identity would itself be vastly improbable). But that then raises the question how those several molecular machines can come together so that proteins from one molecular machine adapt to proteins from another molecular machine to form an integrated functional system like the flagellum. (n.p.)

The answer, which Dembski has missed, is that the flagellum arises in stages.

Rather than (as he implies) a number of subsystems (his "molecular machines") coming together all at once to make a flagellum, a few subsystems came together to make slightly more complex but functional intermediate systems, to which other subsystems were added to make an even more complex but functional intermediate, until finally a primitive motility system that could evolve into the modern flagellum was produced. As we have seen for the archaebacterial flagellum, a swimming flagellum is not suddenly assembled in one go from a secretory system and other bits lying around. The archaebacterial flagellum passed from being a secretory structure, to a gliding-motility system, to a rotatory swimming system. At each point, there was time for substructures to adapt to each other before the next stage.

Dembski is dismissive of type-III secretory systems for another reason. Modern type-III systems are specialized for attacking eukaryotes. Because eukaryotes are supposed to have arisen after flagella, he claims that the type-III systems cannot be ancestral to flagella. But no one has suggested that eubacterial flagella arose from modern type-III systems. Dembski seems unable to contemplate a general, ancestral type-III secretory system, which later specialized into motility and predation systems. Furthermore, many eubacterial flagella are also specialized for attacking eukaryotes, and we do not suppose this means that they arose after the eukaryotes did. Interestingly, predation may be a very old adaptation: some bacteria prey on other bacteria using a hollow pilus not unlike the flagellar filament, so type-III systems may have been involved in predation long before the rise of eukaryotes (Guerrero et al. 1986).

Proposed Evolutionary Pathway

Here is a possible scenario for the evolution of the eubacterial flagellum: a secretory system arose first, based around the SMC rod- and pore-forming complex, which was the common ancestor of the type-III secretory system and the flagellar system. Association of an ion pump (which later became the motor protein) to this structure improved secretion. Even today, the motor proteins, part of a family of secretion-driving proteins, can freely dissociate and reassociate with the flagellar structure. The rod- and pore-forming complex may even have rotated at this stage, as it does in some gliding-motility systems. The protoflagellar filament arose next as part of the protein-secretion structure (compare the *Pseudomonas* pilus, the *Salmonella* filamentous appendages, and the *E. coli* filamentous structures). Gliding-twitching motility arose at this stage or later and was then refined into swimming motility. Regulation and switching can be added later, because there are modern eubacteria that lack these attributes but function well in their environments (Shah and Sockett 1995). At every stage there is a benefit to the changes in the structure.

Dembski may deride this scenario as a just-so story, but we have evidence for it in the form of a variety of intermediates that function well:

1. Simple secretory systems powered by proton motors
2. Gliding-secretory systems powered by proton motors homologous to those of the simple secretory systems, guided by chemical-sensing systems
3. Rotating swimming-secretory systems powered by proton motors homologous to those of the simple secretory systems, guided by chemical-sensing systems homologous to those of the gliding-secretory systems

Thus, we see how a swimming system could arise in stepwise fashion. We also know several other systems that form plausible intermediates: rotating secretory systems and nonrotatory secretory systems with flagellumlike whips. This model is reinforced when we look at other secretory-swimming systems. We have examples of the gliding and swimming cyanobacteria that use modified versions of the same secretory systems. Finally, we have links between type-II secretion, type-IV gliding motility, and archaebacterial flagellar swimming motility to support our model of how eubacterial flagella arose. This evidence shows us that flagella are not isolated swimming machines but one end of a continuum of secretion-motility systems.

Dembski (2003) scathingly says that, in the six years since Behe first claimed that the eubacterial flagellum was IC, researchers have no more than the type-III secretory system to point to. As we have seen, this claim is wholly incorrect. In these years, we have identified yet more homologies between flagellar proteins and other systems, including the critical motor proteins; understood that the archaebacterial and eubacterial flagella are entirely different; and uncovered the deep links between secretion and motility. Given that it has taken nearly 200 years to even begin to understand motility in bacteria, it is amusing that Dembski can declare evolutionary description of the eubacterial flagellum to be a failed project because it has not provided an account that he regards as sufficiently detailed in a mere six years, especially when he is unaware of key knowledge. Indeed, this episode shows clearly how sensitive Dembski's explanatory filter is to background knowledge (which Dembski calls side information; see chapter 8 of this book).

Dembski has claimed that, because the eubacterial flagellum is irreducibly complex, he can eliminate explanations based on natural processes for the origin of the flagellum. This conclusion is wrong for two reasons:

1. Being IC does not eliminate indirect evolutionary explanations (see chapter 5), and flagella can evolve from simpler systems through a series of functional intermediates.
2. Eubacterial flagella are not the outboard motors that Dembski envisages but are organelles involved in swimming, gliding motility, attachment, and secretion. They occupy one end of a range of secretion-based motility systems in bacteria of varying complexity, and several existing intermediate stages show how the flagellum could well have arisen by evolution and natural selection.

Chapter 7

Self-Organization and the Origin of Complexity

NIALL SHANKS AND ISTVAN KARSAI

Even a casual examination of nature reveals the existence of complex, organized states of matter. Organization is found on all scales—for example, in the elaborate spiral shapes of galaxies in space and hurricanes on earth, in organisms, in snowflakes, and in the molecules that participate in many important chemical reactions. Ordered, organized, complex states of matter abound in the world around us. How are we to explain this complexity? Our current best account of these types of phenomena is given by dynamical systems theory, a branch of natural science that explains the existence of complex, organized systems in terms of self-organization.

But natural science has critics who want to explain the existence of organized complex systems as the result of intelligent design by a supernatural being. One such critic is William Dembski (2002b), who modestly claims to have discovered a fourth law of thermodynamics, which he calls the law of conservation of information (169). As Dembski observes, "intelligent design is just the Logos theology of John's Gospel restated in the idiom of information theory" (192). To understand the proposed law, we must see what Dembski means when he refers to what he calls complex specified information (CSI; see also chapter 9 in this book).

Specified events are those forming part of a pattern that can be specified independently of the events. Suppose you want to impress your friends with your skill at archery. You shoot from a distance of 50 meters. Having hit the wall of the barn with all your arrows, you then go and paint bull's-eyes around them and call your friends over to have a look. What can your friends conclude when they arrive at the barn? Dembski (2002b) tells us:

85

> Absolutely nothing about the archer's ability as an archer. Yes, a pattern is being matched, but it is a pattern fixed only after the arrow has been shot. The pattern is thus purely ad hoc.
>
> But suppose instead the archer paints a fixed target on the wall and then shoots at it. Suppose the archer shoots a hundred arrows, and each time hits a perfect bull's-eye. What can be concluded from this second scenario? Confronted with this second scenario we are obligated to infer that here is a world class archer, one whose shots cannot legitimately be referred to luck, but must rather be referred to the archer's skill and mastery. Skill and mastery are of course instances of design. (180)

An archer who draws bull's-eyes around his arrows might generate a pattern, but it won't be a specified pattern. An archer who shoots once and hits the bull's-eye might have been lucky; it could have happened by chance. By contrast, an archer who shoots numerous arrows from a distance and scores many bull's-eyes will have generated a complex, specified pattern of events. Complexity here simply means that the events have a very low probability of occurring just by chance. Dembski claims that when a pattern exhibits complexity and specification and moreover is contingent (that is, is not simply the result of an automatic pattern-generating mechanism), it reveals the presence of intelligent design.

According to Dembski (1999), the law of conservation of information is captured by the claim that natural causes cannot generate CSI. He lays out its implications:

> Among its immediate corollaries are the following: (1) The CSI in a closed system of natural causes remains constant or decreases. (2) CSI cannot be generated spontaneously, originate endogenously or organize itself (as these terms are used in origins of life research). (3) The CSI in a closed system of natural causes either has been in the system eternally or was at some point added exogenously (implying that the system, though now closed, was not always closed). (4) In particular any closed system of natural causes that is also of finite duration received whatever CSI it contains before it became a closed system. (170)

Bringing out a connection with thermodynamics, he observes:

> Moreover, it tells us that when CSI is given over to natural causes it either remains unchanged (in which case the information is conserved) or disintegrates (in which case information diminishes). For instance, the best that can happen to a book on a library shelf is that it remains as it was when originally published and thus preserves the

CSI inherent in the text. Over time, however, what usually happens is that a book gets old, pages fall apart, and the information on the pages disintegrates. The law of conservation of information is therefore more like a thermodynamic law governing entropy, with the focus on degradation rather than conservation. (2002b, 161–62)

What is the connection between Dembski's law and the second law? Dembski's proposed law is related to the second law of thermodynamics through the relationship of information to entropy. He, in fact, asks

whether information appropriately conceived can be regarded as inverse to entropy and whether a law governing information might correspondingly parallel the second law of thermodynamics, which governs entropy. Given the previous exposition it will come as no shock that my answer to both questions is yes, with the appropriate form of information being complex specified information and the parallel law being the law of conservation of information. (166–67)

So he is arguing that as the entropy of a system decreases, information increases, and as entropy increases, information decreases. Any increases in information in a universe such as our own arise from the input of an intelligent designer.

Thermodynamics, Entropy, and Disorder

One of the great achievements of physics in the late nineteenth century was the forging of connections between the basic ideas of thermodynamics and basic ideas of atomic theory, according to which the familiar objects of everyday experience are actually vast conglomerations of tiny particles in jostling motion. Thinking along these lines, let us examine some basic thermodynamical ideas.

Imagine a system that has no exchanges with its surrounding environment (perhaps a large impenetrable box containing cold air and a smallish lump of very hot iron). Such a system is an example of what physicists call a closed system. The first law of thermodynamics, also known as the law of conservation of energy, tells us that the energy of such a system remains constant over time. But even though energy cannot be created or destroyed, it can be redistributed. The second law tells us that the entropy of the closed system will approach a maximum. In practice, the lump of iron will get colder and the air will get warmer until they reach the same temperature. How does this process happen?

Part of the answer is that macroscopic systems like lumps of iron are made of particles. Particles carry energy, and energy is dispersed when particles change their locations by moving about in space, or when energy is transferred from particle to particle as they jostle and bump into each other. The hotter macroscopic systems are, the more energy their constituent particles have and hence the more vigorously these particles move and jostle. The iron in the box cools and the air warms because particles of iron jostle particles of air, thereby transferring energy to them. In this way heat flows from the hotter to the cooler. Energy is redistributed from the iron to the air until the iron is at the same temperature as the air, at which point there is no net energy flow between them.

To better understand the significance of seeing heat in terms of the motions of particles, we will differentiate between coherent motions of particles and random, incoherent, thermal motions of particles (Atkins 1994). A gas stove takes the chemical energy in gas and converts it into heat energy. When gas burns, energy disperses through incoherent, random motions of particles. These particles jostle particles in the pan on the stove, which disperse energy by transferring it to the water molecules in the pan. As these jostle faster, the water gets hot, and you can make tea.

By contrast, consider a car. When a piston in a cylinder goes up and down, there is a net movement up and down of the particles making up the piston as well. These are coherent motions. When we get work from such a system, it is because we are able to use energy to induce and sustain coherent motions of the particles making up the car engine. Thus, coherent motions in one part (the reciprocating motion of the pistons in the engine block) are converted through coherent motions in other parts (cranks and gears) into coherent, rotary motions of the wheels. In virtue of these coupled, coherent motions, by burning gasoline, you can drive yourself to the store to buy tea.

Cars work because they are physical systems whose parts (made of tiny particles) stand in appropriate structural relationships to each other so that coherent motions in one part can cause appropriate coherent motions in other parts. But even the best cars are subject to thermal wear and tear. As the chemical energy in gasoline is consumed to run the car, frictional heating causes brake pads to wear out. Electrical heating wears out spark plugs. Metals get fatigued (structural changes occur as particles vibrate and change locations), and parts drop off. As the particles that make up the car's parts change location and jostle each other, the car gradually loses its structural coherence and eventually suffers catastrophic failure. This is what it means to say that the entropy of the car increases over time.

How, then, can self-organization possibly occur? How can natural mechanisms operating in accord with the laws of nature bring about entropy reduction and give rise to order and information without the intervention of an intelligent designer to both organize things and keep them organized?

Self-Organization and the Emergence of Order

To find out how order forms, we must distinguish between closed systems, which have no exchanges with their surrounding environments, and open-dissipative systems, which have such exchanges. Our universe contains many open-dissipative systems. When energy and matter flow into and out of open-dissipative systems, they can drive the formation and maintenance of coherent structures and coherent dynamical processes in systems by inducing coherent, coordinated motions in matter—that is, in atoms and molecules. The processes by which these coordinated motions of matter are induced are known as self-organizing processes (since they involve no external designing agency). The complex organization that results from these processes is generated by energy-driven interactions among the components internal to open-dissipative systems.

In accord with the first law, the law of conservation of energy, the work involved in the formation and maintenance of coherent structures in open-dissipative systems happens as a result of energy flowing through the system. Nature does not give something for nothing, and there is no energetic free lunch. The entropy reduction involved in the formation and maintenance of coherent structures and processes must, in accord with the second law, be more than offset by an increase in the entropy of the environment with which it interacts. This last statement means that the formation and maintenance of coherent structures and processes involve the corruption of usable energy in the universe, where the universe is a system currently teeming with usable energy.

To get self-organization, several conditions need to be satisfied. These include the following.

A COLLECTION OF SUITABLE COMPONENTS
The components come in all shapes and sizes. They can be of differing kinds. They may be atoms or molecules (water will do); they may be cells; they may be organisms (for example, an insect in an insect society); they may even be the stellar components of galaxies self-organizing through gravitational energy into giant rotating spirals.

A FLOW OF USABLE ENERGY THROUGH THE SYSTEM
This flow of energy drives mechanisms that give rise to the formation of self-organizing systems. The flow of energy into and out of the system must continue

to sustain the system by driving interactions among its components. A self-organizing system starved of sustaining energy will sink back into the environment from which it emerged as it loses its structural and dynamical coherence. The flow of warm, moist air that drives the formation and sustenance of large self-organized structures such as hurricanes (visible from space as rotating spirals) is disrupted by landfall, whereupon the weather system settles down, spawning self-organizing tornadoes in its death throes. For self-organized creatures like us, as the great nineteenth-century physiologist Claude Bernard was among the first to emphasize, equilibrium is death.

LOCAL COUPLING MECHANISMS

The components must be able to couple their behaviors in accord with local mechanisms. The locality condition means that interactions giving rise to self-organization take place between proximate components of a system with no broader view to the integrated, complex system than may result from many such purely local interactions. The integration of the components of self-organizing systems into organized, complex systems arises from chains of local interactions (as when ants interact with each other through intermediaries—possibly other ants, possibly a pheromone trail laid down by an ant no longer present). This coupling of the behaviors of the components lies at the heart of self-organization. Self-organizing systems are systems of many interacting parts whose interactions with each other give rise to the global, collective behavior of the entire system of interacting components.

For example, a self-organized structure such as a hurricane has air molecules and water molecules as components. It is an entity whose formation is driven by heat energy flowing from the ocean to the upper atmosphere. A hurricane begins with a tropical depression (a point of low air pressure) that draws in warm moist air from the immediate surroundings. The water vapor in the air condenses and falls as rain as the air is drawn into the region of low pressure. As water changes from vapor to liquid, it releases energy as latent heat. This heat causes the air to rise at the center of the emerging structure, helping to form the eye of the storm, thereby drawing in more warm, moist air from below and ultimately from outside the system. This air in turn surrenders its water vapor and rises up the wall of the eye. The resulting coordination of air flows contributes to the emergence of global behaviors of the entire system: organized, rotating, spiral patterns that can be seen from space.

The local coupling of components (in our hurricane, air rising up the wall of the eye drawing in more air from below, which in turn draws in air from outside the system) constrains their behavior and thus their freedom to respond to changes in their immediate environments. This feature of self-organizing systems is important for understanding how energy flows through a system can induce the coordinated, coherent motions that result in the organization exhibited by such systems. (The local coupling of components also influences how environmental influences will usually be able to propagate through the system. The extent of the propagation will depend on the presence or absence of amplification mechanisms, damping mechanisms, and other factors—for example, how tightly the components are coupled.)

The dynamical stability of self-organizing systems is due to regulatory mechanisms. Positive feedback will make a system grow by amplifying initial effects. An important positive-feedback mechanism is autocatalysis, where the very presence of something, given a source of usable energy, promotes the formation of more of itself. Autocatalysis plays important roles in physics, chemistry, and biology (Shanks 2001). For a simple example, take rabbits and add grass for energy. The result, in the fullness of developmental time, is more rabbits than you began with.

But we do not see arbitrary, uncontrolled growth in the rabbit population, so positive feedback must be balanced by negative feedback. A growing rabbit population, for example, draws the attention of foxes, who eat the rabbits—and produce more foxes in consequence. The rising fox population leads to overpredation of the rabbit population, and this in turn causes the rabbit population to collapse. With a diminished food source, the fox population will shortly collapse and enable the rabbit population to grow again. The result, over time, will be two coupled populations, whose numbers will rise and fall together. We will in fact have a biological oscillator.

Bénard-Cell Convection

Consider a thin layer of water sandwiched between two horizontal glass plates. Suppose the system is at room temperature and in thermal equilibrium with its surroundings. One region of water looks pretty much the same as any other. If the water is now warmed from below so that energy is allowed to flow through the system and back into the environment above, the system will become self-organized above a certain critical temperature. If you look down at the system, you will see a structured, honeycomb pattern in the water (see figure 7.1).

The cells in the honeycomb—often shaped like hexagons or pentagons—are known as Bénard cells and are rotating convection cells. Water warmed at the bottom rises; as it rises it cools and starts to sink again to the bottom to be rewarmed, thereby repeating the process. Water cannot both rise and fall in the same place, so regions where water rises become differentiated from

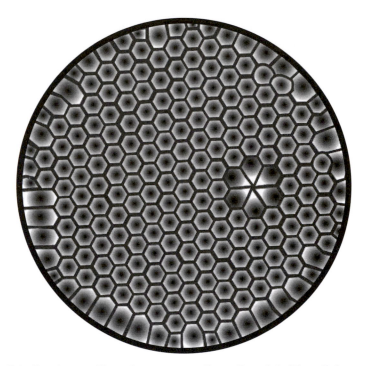

Figure 7.1. Simulation of Bénard convection cells in a Petri dish. The cells have similar size, except on the border. Although the shapes of the cells vary somewhat, they approach a hexagonal structure commonly called honeycomb. The emergence of an organized structure from a homogeneous medium such as water or oil is startling.

regions where it falls. This differentiation gives rise to the cells. Seen from the top, the cells have a dimpled appearance, since water rises up the walls of the cell and flows toward the center dimple to flow back down again, completing the convective circulation (see figure 7.2).

The cells are visible because of the effects of temperature on the refraction of light. The way in which one cell rotates influences the ways in which its immediate neighbors rotate; in turn, the first cell is influenced by them. By adding thermal energy to water, we have brought about the spontaneous emergence of a complex system of interacting convection cells. The spatial and temporal order we can see in the behavior of this self-organizing system is not imposed from outside. The environment merely provides the energy to run the process. Chance, in the form of environmental fluctuations, provides the initial local inhomogeneities that serve as seeds for the emergence of the system from an initially homogeneous aqueous medium. The Bénard-cell patterns result from the energy-driven interactions of the components (water molecules) internal to the system. Bénard cells are not just an artificial phenomenon: astronomers have seen these cells on the surface of the sun.

Self-organizing systems, such as the Bénard-cell system, constitute a threat to Dembski's creationist enterprise because, although these systems are both undesigned and naturalistically explicable, they manifest complex specified information and thereby give the misleading appearance of being the fruits of intelligent design.

Apparently aware of the threat posed by self-organization of this kind to his claims about intelligent design, Dembski (2002b) initially accuses those who study these phenomena of trying to get a free lunch:

> Bargains are all fine and good, and if you can get something for nothing, go for it. But there is an alternative tendency in science that says that you get what you pay for and that at the end of the day there has to be an accounting of the books. Some areas of science are open to bargain-hunting and some are not. Self-organizing complex systems, for instance, are a great place for scientific bargain-hunters to shop. Bénard-cell convection, Belousov-Zhabotinsky reactions, and a host of other self-organizing systems offer complex organized structures apparently for free. But there are other areas of science that frown on bargain-hunting. The conservation laws of physics, for instance, allow no bargains. (23)

Yet Bénard cells occur in nature (for example, in the sun) as well as in the laboratory. Their existence is certainly consistent with known conservation laws.

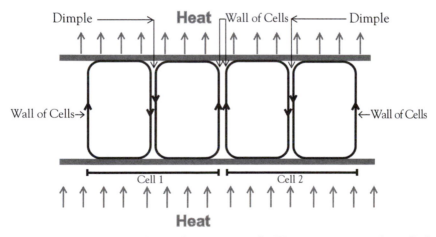

Figure 7.2. Cross-section of Bénard convection cells. Warm water rises up the wall of the cell, cools, and sinks down at the dimple in the middle of the cell.

The matter is made all the murkier because Dembski says elsewhere that he finds the existence of Bénard cells to be unproblematic (which seems to contradict his suggestion that they violate the conservation laws of physics). Thus, he observes:

> Bénard-cell convection, for instance, happens repeatedly and reliably so long as the appropriate fluid is sufficiently heated in the appropriate vessel. We may not understand what it is about the properties of the fluid that makes it organize into hexagonal cells, but the causal antecedents that produce the hexagonal patterns are clearly specified. So long as we have causal specificity, emergence is a perfectly legitimate concept. (243)

Here Dembski is guilty of gross oversimplification in his attempt at an easy rebuttal of a difficult problem.

What actually happens "repeatedly and reliably" is a pattern involving some arrangement of rotating convection cells (often involving both hexagons and pentagons), where the rotation of one cell reflects and is in turn reflected by the rotations of its neighbors. But we do not get the same pattern (including rotation dynamics) each time we run the experiment. The actual pattern generated in a given trial reflects both the Bénard-cell convection mechanism and the effects of chance inhomogeneities and fluctuations in the fluid medium. In consequence, the precise patterns generated in a sequence of trials exhibit a high degree of variation. These contingent patterns are nothing like the results of an automatic pattern-generating mechanism that gives the same result repeatedly and reliably, time after time: the Bénard-cell patterns also exhibit complex specified information.

To see this, consider once again the patterns generated by Dembski's archer, who intelligently and skillfully designs the trajectories of his arrows to hit the bull's-eye of a target from a great distance. A pattern of several hits in the bull's-eye is complex because it has a low probability of happening by chance alone. The general form of the pattern—a pattern involving hits in the region of the bull's-eye—can be specified in advance (or independently) of the shooting of the arrows. That the pattern of hits is skillful and not the result of an automatic pattern-generating mechanism is manifested in the observation that, whenever the archer shoots several arrows to demonstrate his skill, he does not repeatedly and reliably get exactly the same pattern of hits in the region of the bull's-eye. The actual patterns of hits generated in a sequence of trials are contingent.

Bénard-cell patterns are complex: they involve the coordinated motions of trillions of water molecules, and the probability that they would form by chance alone is minuscule. As with the archer, the general form of the pattern, involving some arrangement of rotating hexagons and pentagons, is specifiable in advance (and independently) of any given trial. As with the archer, the actual pattern generated on any given trial is contingent. You do not get exactly the same pattern repeatedly and reliably each time you run the experiment. The crucial difference between the Bénard-cell pattern and the archer's pattern of hits is that the Bénard-cell pattern does not require intelligent design or skillful manipulation for its appearance, only the combined effects of a dumb pattern-generating mechanism and mindless chance in the form of fluctuations and inhomogeneities in the fluid medium.

The problems posed by Bénard-cell patterns for intelligent-design theorists such as Dembski do not end here. As we saw at the beginning of this chapter, Dembski claims that information is the inverse of entropy. The emergence of the Bénard-cell patterns involves a local decrease in entropy (that is, a decrease of disorder or an increase of order). It follows from Dembski's claim that, when Bénard cells form, as entropy decreases, information increases. But this increase of information does not involve any input or use of complex specified information arising from intelligent causes, be they natural or supernatural. All that is needed are unintelligent, natural mechanisms operating in accord with the laws of physics.

Dembski claims that his law of conservation of specified information precludes the formation of complex systems through natural causes. But the universe we live in has lots of usable energy and is far from thermodynamical equilibrium. (At equilibrium both entropy and information would remain constant, on average.) The universe also contains many open-dissipative systems as subsystems. For example, our planet is warmed by a large hot star that provides plenty of usable energy, and the universe, not to mention our planet, is teeming with open systems that exploit this usable energy. Self-organization, resulting in decreases in the entropy of local, open systems, points clearly to the conclusion that we can indeed get CSI through self-organization resulting from unintelligent natural causes and that no invisible supernatural hand operating outside a system of purely natural causes is needed. Self-organization is indeed a great scientific bargain when compared with evidentially empty promissory notes concerning supernatural design from outside our natural universe.

Self-Organization As a Pathway to Irreducible Complexity

Michael Behe, a creationist biochemist and a leading light in the intelligent-design movement, has argued that there is a kind of complexity in nature called irreducible complexity that can exist only as the result of the activity of an intelligent designer (see chapter 4 in this book). An irreducibly complex system is one consisting of several components, all of which must be present if the system as a whole is to achieve its function. Dembski (1999, 149) has attempted to bolster these claims by arguing that irreducibly complex systems also manifest complex specified information, which, as we have seen, he takes as the hallmark of intelligent design.

Behe has illustrated his idea of irreducible complexity with the example of a mousetrap (1996; 2000; 2001a, 90–101; also see chapter 2 in this book). The mousetrap is a device that has several components, all of which are necessary to catch mice. Assume for the sake of argument that Behe is right about all this. He tells us that, although it is easy to see how such a complex, structured system could arise by intelligent design and construction (it is, after all, a human artifact), it is hard to see how it could have formed through the operation of unintelligent, natural mechanisms. The components of mousetraps will not self-assemble into a functioning mousetrap. Yet Behe intends the mousetrap to serve as a metaphor to illustrate the complexity of chemical reactions. It is far from obvious that chemical reactions with the property of irreducible complexity necessarily result from intelligent design.

In chemistry, self-assembly and self-organization are well-studied phenomena. One of the most famous and well-studied self-organizing chemical systems is the Belousov-Zhabotinski (BZ) reaction. The BZ reaction refers to a set of chemical reactions in which an organic substrate is oxidized in the presence of acid by bromate ions in the presence of a transition metal ion (Tyson 1994, 569–87).

The version of the reaction that one of us (Niall Shanks) has used in classroom demonstrations has the following ingredients: potassium bromate, malonic acid, potassium bromide, cerium ammonium nitrate, and sulfuric acid. When the ingredients are placed in a beaker, the system self-organizes to perform a repeating cycle of reactions. It behaves as a chemical oscillator, and the oscillations can be monitored through cycles of color changes. You can use it to tell the time: it is a watch that forms in a beaker without the help of a watchmaker.

The oscillations result from the chemical system cycling through its component reaction pathways. What do we mean? Suppose the system starts out with a high concentration of bromide ions. In the first group of reactions, bromate and malonic acid are used in a slow reaction to produce bromomalonic acid and water. Bromous acid is one of the reaction intermediates in this pathway. Since the cerium is in the cerous state, the reaction medium remains colorless for this phase of the cycle. As time goes by, the concentration of bromide ions drops to a point at which bromous acid can initiate another mechanism to produce bromomalonic acid and water.

Here, in a fast reaction, bromate, malonic acid, bromous acid (a reaction intermediate from the first pathway), and cerous ions produce ceric ions, bromomalonic acid, and water. The reaction medium turns yellow as cerium enters the ceric state. The pathway also contains an autocatalytic step in which the very presence of bromous acid catalyzes the production of more of itself, so one molecule of bromous acid makes two molecules of bromous acid (this positive feedback effect is why this pathway is fast). As cerous ions are consumed and ceric ions accumulate, a critical threshold is achieved, and a third pathway opens. This pathway consumes bromomalonic acid, malonic acid, and ceric ions to produce carbon dioxide and bromide ions, and to regenerate cerous ions, thereby setting the system up for a new cycle (Babloyantz 1986).

Neither the law of conservation of energy nor the second law is violated. To get the oscillations, the system begins far from chemical equilibrium. The oscillations continue until equilibrium is reached: the period gradually gets longer and the color changes become less pronounced as equilibrium is approached. Like more familiar mechanical watches, it runs down unless it is rewound by the addition of more reagents. We have had the system oscillate for more than an hour in typical classroom demonstrations. That the reaction manifests self-organization means nothing more than that the invisible hand of the chemical interactions between molecules, in accord with the laws of chemistry, brings about highly ordered, organized behavior of the system as a whole in the form of regular temporal oscillations. This behavior does not require the intervention of a supernatural intelligence.

The reaction is important because advocates of intelligent-design theory claim that irreducible complexity can appear only as the result of the actions of an intelligent designer who takes the components of the system and assembles them into a functioning whole. In saying this, they evidently mean that they cannot see how unguided mechanisms operating in accord with the laws of nature could give rise to this type of complexity. But the BZ system manifests irreducible complexity, and it does so without any help from intelligent designers (Shanks and Joplin 1999, 2001; Shanks 2001). How can this be so?

Behe (1996, 2000) tells us that three conditions must be satisfied if a system is to be irreducibly complex: (1) the system must have a function, (2) the system must consist of several components, and (3) all the components must be required for the achievement of function. The function of the BZ reaction is to oscillate. The BZ system consists of several key reactions. The key components of the BZ reaction are all needed for the oscillatory cycle to exist. The disruption of any of these key reactions results in the catastrophic failure of the system. The BZ system manifests the same irreducible complexity found in a mousetrap, yet it requires no intelligent designer to arrange the parts into a functioning whole. Apparently, the unguided laws of chemistry will generate irreducibly complex systems.

Yet Behe (2000) has objected to this example. He observes, "Although it does have interacting parts that are required for the reaction, the system lacks the crucial feature—the components are not well-matched" (157). This charge has been reiterated by Dembski (2002b), who tells us that being well-matched means being like the fan belt of a car: "specifically adapted to the cooling fan" (283). Behe (2000) thus objects that the reagents used in the BZ reaction have a wide variety of uses. In his terminology, they have low specificity (158). For example, one ingredient, sodium bromate, is a general-purpose oxidizing agent; and ingredients other than those we mentioned can be substituted. As we have noted, the term BZ *reaction* refers to a family of chemical reactions.

If Behe is right, then mousetraps are not irreducibly complex either. Their components also have low specificity. The steel used in their construction has a wide range of uses, as does the wood used for the base. You can substitute plastic for wood and any number of metals for the spring and hammer. Mousetraps are easy to make (which is why they are cheap) and will work with metals manifesting a wide range of tensile strengths. Either the BZ system is an irreducibly complex system, or the mousetrap is not a model for irreducible complexity. Take your pick, because you cannot have it both ways.

Moreover, crucial components of Behe's own biochemical examples of irreducible complexity have multiple uses and lack substrate specificity (interact with a wide variety of substrates). For example, plasminogen (a component of the irreducibly complex blood-clotting cascade) has been documented to play a role in a wide variety of physiological processes, ranging from tissue remodeling, cell migration, embryonic development, and angiogenesis as well as wound healing (Bugge et al. 1996). And although Behe (1996) tells us that plasmin (the activated form of plasminogen) "acts as scissors specifically to cut up fibrin clots" (88), we learn in one of the very papers he cites that "plasmin has a relatively low substrate specificity and is known to degrade several common extracellular-matrix glycoproteins in vitro" (Bugge et al. 1996, 709). This component of an irreducibly complex system is evidently nothing like the fan belt of a car "specifically adapted to the cooling fan."

Nor, for that matter, are all the components of the clotting pathway necessary for function. Plasminogen-deficient (Plg–/–) mice (hence, mice lacking plasmin) have been studied. As noted, plasmin is needed for clot degradation, yet as Bugge at al. (1996) comment,

> Plasmin is probably one member of a team of carefully regulated and specialized matrix-degrading enzymes, including serine-, metallo-, and other classes of proteases, which together serve in matrix remodeling and cellular reorganization of wound fields. . . . However, despite slow progress in wound repair, wounds in Plg–/– mice eventually resolve with an outcome that is generally comparable to that of control mice. Thus an interesting and unresolved question is what protease(s) contributes to fibrin clearance in the absence of Plg? (717)

The reasonable conclusion is that, if Behe's examples are indeed examples of irreducibly complex systems, then so is the BZ system. Hence, self-organization is evidently a pathway to irreducible complexity and one that involves no intelligent design, supernatural or otherwise.

Construction without Intelligence

Looking at the pyramids of Giza, we see huge, intelligently designed, complicated structures built by humans about 4000 years ago. These are structures with a definite function. They are not natural formations; thousands of people built them over many decades. The work was carefully planned and executed. The structure is a result of the planning of architects, the blueprints of engineers, the organization of bureaucratic and military commanders, and the work of many laborers.

Social wasps construct paper nests with complexity and relative size that is similar to that seen in structures of intelligent human construction. Where are the blueprints, the engineers, and the hierarchical chain of command in the execution? As Maurice Maeterlinck (1927) asked, "What is it that governs here? What is it that issues orders, foresees future, elaborates plans and preserves equilibrium?" (137)

These are interesting questions, because the structure built by insects with their tiny brains and limited intelligence seems to be beautifully regular and complicated even for us human beings. It seems certain, however, that no wasp possesses knowledge of the ultimate form of the structure, and the duration of the building process generally spans several lifetimes of an individual. Apparently, coordinated construction activity does not depend on supervisors. As biologist Thomas Seeley (2002) has observed, "The biblical King Solomon was correct when he noted [in Proverbs 6:7], in reference to ant colonies, there is no guide, overseer or ruler" (315).

Organizing construction activity among humans generally requires a well-informed leader who directs the building activity of the group, providing each group member with detailed instructions about what to do to contribute to the global result. A group of unskilled workers building a barn under the command of a master carpenter is a good example. The resulting barn is not the result of self-organization, because we can halt the construction by removing the master carpenter or by blocking the information flow from the master carpenter to the other workers.

Construction of more-sophisticated structures generally requires something more than a construction leader. These activities also require blueprints. Blueprints are compact representations of the spatiotemporal relationships of the parts of structures. A blueprint may be a small replica (a scale model) or a detailed drawing, perhaps accompanied by explanations.

Blueprints result from the creative acts of intelligent designers. They typically require skilled on-site interpretation. They also enable the construction workers to produce a more-sophisticated structure than they could without the blueprint. Simply following a set of instructions is similar in several respects to using a blueprint. A set of instructions provides step-by-step construction procedures that typically do not require skilled interpretation. A good example is the construction of an elaborate Lego structure, following the directions of an enclosed booklet that shows which kinds of blocks have to be attached to the incipient structure and in what order. None of these approaches to the construction of structures is based on self-organization. Removing the blueprint or the set of instructions will stop the construction or lead to disaster.

Humans also use templates to construct simple items. Templates are different from blueprints and sets of detailed instructions: rather than functioning as an aid for workers to carry out complicated construction, templates ensure the production of consistent and reproducible units such as bricks. There are numerous analogs of these human approaches to the design and construction of structures in the nonhuman, biological world (Camazine et al. 2001).

Social wasps build nests to keep their carnivorous larvae in one location yet separate them from each other. Early analysis of construction behavior involved little more than division of the behavior into acts of instinct and acts of intelligence. Thorpe (1963), for example, went so far as to claim that wasps use a mental blueprint to guide nest construction. Results of further experiments and perturbation of the construction behavior suggested that, instead of a mental image, the construction behavior was driven by an inherited building program: a set of instructions coded in the wasps' genes (Evans 1966).

The main problem with these early ideas was that they did not include any analysis of the role of ongoing inspection of the changing state or condition of the incipient structure and the subsequent use of this information to modify the behavioral states of the insects involved in nest construction. It was also assumed that "the more complex the nest construction becomes . . . the more sophisticated the building programme must be. Hierarchical level of evaluation, subroutines within the building programme, and learning capabilities appear to be the ways of achieving this sophistication" (Downing and Jeanne 1990, 105). Learning, along with use of construction leaders, blueprints, and sets of instructions, is costly and may require developed cognitive abilities. With the possible exception of learning, these other approaches to construction are often highly sensitive to small errors whose consequences can rapidly become catastrophic.

In fact, it now looks as though social insects rely on simple self-organizing construction processes that do not require sophisticated cognitive abilities and are also error-tolerant. The explanation we will provide here is based on decentralized coordination, in which individuals respond to stimuli provided through the common medium of the emergent nest. In the case of the collective building of a wasp nest, where many individuals contribute to the construction, stimuli provided by the emerging structure itself can be a rich source of information for a given individual. The term *stigmergy* (Grassé 1959) describes the situation in which the product of previously accomplished work, rather than direct communication among the builders, induces the wasps to perform additional labor.

In a stigmergic account of nest construction, the completed nest is a complex structure whose specifiable morphology reflects the behavioral repertoires of the insect builders as they respond to a multiplicity of chance encounters with a changeable, contingent environment during the construction. The construction is thus not the unfolding of a preordained plan—intelligent, genetic, or otherwise. As environmental encounters vary, so do the shapes of the nests constructed. Construction is not teleological; it occurs with no view to the future.

Moreover, even when two nests have more or less the same shape, they are not built in exactly the same way, repeatedly and reliably, as if by an automatic, preprogrammed process. Stigmergic accounts of nest construction recognize that there are many construction pathways to a nest of a given general shape. The pathway actually taken (which wasp does what in response to local cues and when) reflects both the internal states of the wasps and the many chancy, unpredictable contingencies associated with the actual construction.

The result of this dumb process is a complex structure that gives the misleading appearance of being intelligently designed. Here is how it happens.

Hexagonal cells are the basic unit of the wasp comb. Hexagonal cells are a very efficient way to fill a two-dimensional space and also very economical. But how, exactly, do these regular structures emerge? Detailed observations show that hexagonal forms are a predictable by-product of cell-building activity and do not require any higher-level rule or information (West Eberhard 1969, Karsai and Theraulaz 1995).

The hexagonal cells emerge from wasps' attempts to make conelike structures. When a wasp lengthens a given cell, it also tries to increase its diameter. Imagine that the wasp builds a cone by adding a small quantity of material to the lower edge of the cone. Several cones are linked, however; and if the wasp detects another cell adjacent to the cell it is building, it slightly modifies its posture and does not extend the cell in that direction. The result of this behavior can be seen very clearly in the cells that are on the periphery of the comb. They have two or three neighbors, and all sides facing outward are curved (see figure 7.3). Later, when new cells are added to the comb, these outer cells become inner cells and are turned into hexagonal cells. The hexagonal shape emerges without a blueprint, as a result of a simple building rule that is based only on local information. The hexagonal cell is just one of the

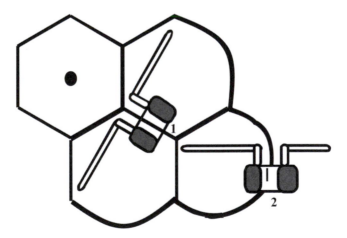

Figure 7.3. Cell shaping by wasps (head shown only, view from below). The cell with the black dot has six neighbors (just two are drawn) and has a perfect hexagonal shape. Peripheral cells have a straight border with their neighbors, but neighborless sides are curved. In these cells, wasp 2 increases the diameter of the cell by pushing the building material outward while its head is tilted. When a cell wall is built between two cells, the head of wasp 1 is not tilted, and the cell wall becomes a straight line.

emergent regular characteristics of the wasp nests. These cells form a comb, which has a definite (generally regular) structure. One of the most common comb shapes is a hexagonally symmetrical shape (see figure 7.4).

The hexagonally symmetrical comb shape has several adaptive advantages: it requires less material per cell, is better in terms of heat insulation, and, because of its small circumference, can be protected easily. But the adaptive explanation of this compact cell arrangement will not tell us how wasps built this structure. Philip Rau (1929) concluded from his experiment that the hexagonal symmetry is learned. Istvan Karsai and Zsoltan Pénzes (1993) analyzed the nest structures and the behavior of wasps and argued that the construction is based on stigmergy.

In a stigmergic type of construction, the key problem is to understand how stimuli are organized in space and time to ensure coherent building. The hexagonally symmetrical structure emerges as a global pattern without deliberate planning. It is a by-product of simple rules of thumb that are triggered on the basis of local information (the wasps do not experience or conceive the shape of the comb).

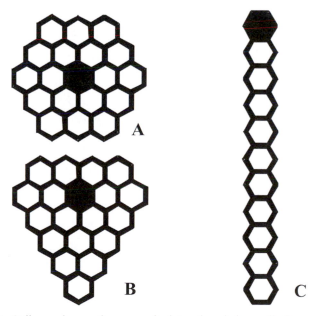

Figure 7.4. Different shapes of wasp combs (view from below, cells drawn as idealized hexagons). Dark cells are the first cell where the comb is attached to the substrate. Type A is the most common hexagonally symmetrical comb; a type-B comb has a single symmetry axis; and type C, the rarest, has a single cell row. The combs can grow much larger while keeping the same form.

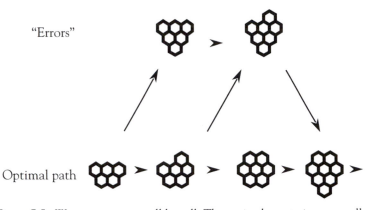

"Errors"

Optimal path

Figure 7.5. Wasp nests grown cell by cell. The optimal way to increase cell number is depicted in the bottom row. In real wasp nests "errors" (suboptimal forms) emerge as well. Faithful use of a blueprint or recipe would not allow errors in the system. The errors indicate that the wasps use a simple rule of thumb to construct their nest. Analysis of these errors helped investigators find the rule of thumb guiding construction.

Karsai and Pénzes (2000) examined several candidate rules of thumb and compared predicted nest forms to natural nest forms. They found that not all of the nest forms in nature have an optimal shape. These suboptimal forms could be explained away as anomalies, or they could be consequences of the rules of thumb (see figure 7.5). Karsai and Pénzes considered these "faulty" nests to be real data and the inevitable consequence of the rule of thumb actually used. The next step in their analysis was to find the rule of thumb that generates all of the optimal shapes as well as the faulty structures (that is, the complete set of natural nests).

Karsai and Pénzes (2000) examined the predictions of several candidate rules of thumb. One of the rules was able to generate the whole set of natural nest forms. The rule can be described in functional terms as follows: construct a new cell, where the summed age of the neighbors of the new cell shows the maximum value (see figure 7.6). This rule gives rise to the maximum age model.

Karsai and Pénzes showed that a beautiful, regular, and adaptive structure emerges even if the builders are unaware of this regularity. The builders follow simple rules. As the nest grows and changes during construction, the nest itself provides new local stimuli to which the rule-following builders respond. As the builders respond to changing local stimuli, a globally ordered structure emerges. It is as if the developing nest governs its own development; the builders are only the tools. The wasps do not follow the ages of cells and sum their ages for their decision. In fact, several parameters correspond to the

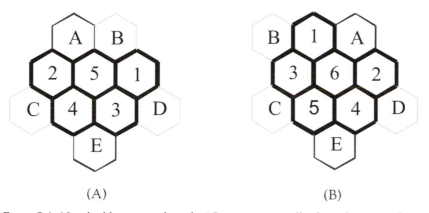

(A)   (B)

Figure 7.6. Nest building using the rule, "Construct new cell, where the summed age of the neighbors of the new cell show maximum value (that is, a maximum age)." Thick cells show the current structure. Numbers in the cells show their age. (In example [a], the upper middle cell with number 5 is the oldest.) Cells with thin walls and letters in them show the locations of initiation positions in case of random cell initiation. In example I, for positions A (2+5) and E (4+3), the summed age of the neighbor is equal; thus, the maximum age model predicts one of two possible forms to emerge. If position A is chosen for the sixth cell, then we have a six-celled form shown in example (b). Here again, the maximum-age model selects positions A and E with the same stimulus strength (9), which means that two possible forms can emerge again.

age of cells: cells become longer and wider as they age, and they absorb more chemicals. These constitute the local information that can be sensed by the wasps.

Now that we have explained how the regular hexagonally symmetrical comb shape emerges, it is natural to try to understand how other comb shapes emerge (see figure 7.4). Does every shape need a unique rule of thumb? Using the stigmergy approach, Karsai and Pénzes (1998) showed that the variability of comb forms can be deduced from the same construction algorithm. Tweaking a single parameter of the model, the authors generated all forms found in nature and, interestingly, only those. This shows that variability and complexity may emerge in a very simple system in which interacting units follow simple rules and make simple decisions based on the contingencies of local information.

Communities of nest-building wasps are open-dissipative systems. The internal dynamics of these systems is driven by flows of energy through the system and constrained by parameters derived from the environment with which the insects interact. The elaborate, structurally coherent nests are highly improbable forms that could not have arisen by chance. In fact, these orderly, low-entropy structures emerge as the products of interactions between the insects that constitute the nest-building community and their immediate environments. These structures require no intelligent design from outside the system, nor do they require a guiding intelligence, be it a single individual or collective of individuals, operating within the system. The orderly, complex structures emerge as the consequence of the operation of blind, unintelligent, natural mechanisms operating in response to chancy, contingent, and unpredictable environments.

Acknowledgments

We are grateful to Taner Edis, Matt Young, Jason Rosenhouse, and Mark Perakh for helpful comments on earlier versions of this chapter.

Chapter 8

The Explanatory Filter, Archaeology, and Forensics

GARY S. HURD

Some years ago, I was invited to visit the new laboratory of a senior forensic anthropologist. After I had admired all the latest features (and been reduced to a puddle of envy), she told me that there was something else she wanted me to see. She left the lab and in a moment returned with a large plastic bag. In the bag was the lower part of a human leg. The flesh was missing from the upper portion of the tibia and fibula, but flesh could still be seen inside a woman's hiking shoe.

I borrowed a microscope to examine some marks in the exposed bones. They were basically identical with those I had examined hundreds of times before. Young canine puppies (most likely coyote) had used these bones for teething and weaning. This placed the woman's time of death sometime in the late spring or early summer. From the condition of the protected soft tissues inside the shoe and the strong smell, I concurred that the death had occurred within the preceding year but concluded that the marks did not bring us any closer to determining a cause of death.

On another occasion, a visitor to the museum where I worked opened up a paper bag and took out a rock. It was roughly triangular on the major axis, lens-shaped on the minor axis, and had a large number of flaking scars over its surface. The visitor asked me if I could tell him what it was, how old it might be, and if it was worth any money. I told him that it was a busted rock that was worth only what someone might give him for it. Many people would have called it an arrowhead.

What do these anecdotes have to do with intelligent-design "theory"? According to William Dembski, I have been executing the same intellectual activities he employs to discover the existence of God.

Mathematician and theologian Dembski is one of a handful of academically trained individuals who advocate intelligent design (ID). The bulk of ID polemic is undistinguished from other forms of creationist opposition to evolutionary biology (Johnson 1993, Wells 2000). Dembski's work and that of biochemist Michael Behe (1996) differ slightly in that they claim to offer empirically substantiated and theoretically rigorous demonstration that an intelligent designer is responsible for specific features of life at the molecular and cellular level.

Dembski (1994) gives us his criteria for the designer:

> I look for three things in a supernatural Designer—intelligence, transcendence and power. By power I mean that the Designer can actually do things to influence the material world—perform miracles if desired. By transcendence I mean that the Designer cannot be identified with any physical process event or entity—the latter can at best be attributed to, not equated with, the Designer. By intelligence I mean that the Designer is capable of performing actions that cannot adequately be explained by appealing to chance—the Designer can act so as to render the chance hypothesis untenable. (116)

J. P. Moreland (1999), professor at the Talbot School of Theology at Biola University (the Bible Institute of Los Angeles), offers this summary of Dembski's (1998c) program:

> William Dembski has reminded us that the emerging Intelligent Design movement has a four pronged approach to defeating naturalism: (1) A scientific/philosophical critique of naturalism; (2) a positive scientific research program (Intelligent Design) for investigating the effects of intelligent causes; (3) rethinking every field of inquiry infected with naturalism and reconceptualizing it in terms of design; (4) development of a theology of nature by relating the intelligence inferred by intelligent design to the God of Scripture. (97)

Dembski insists that there is a sound scientific base for the intelligent-design position, regardless of its supernatural origins. He claims that his method of inquiry is typical of what he calls the "special sciences," including archaeology and forensics, my professional areas. Dembski uses an explanatory filter (EF) to identify design. After a brief discussion of the EF, I will illustrate why these sciences cannot be used to legitimize it.

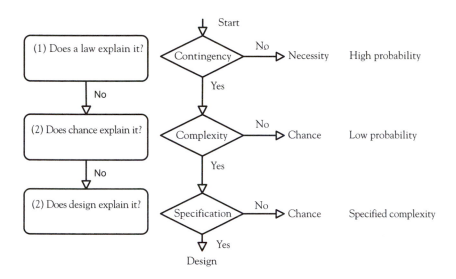

Figure 8.1. A composite representation of the explanatory filter. After Dembski (1998a, 1998b, 1999, 2002b).

The Explanatory Filter

Figure 8.1 represents the EF as it has been variously described by Dembski (1998a, 1998b, 1999, 2002b). It is appropriate to modify a model system in response to criticism or as a reflection of changes in your thought, and Dembski's inconsistencies are not pointed out as criticisms per se. Rather, my intent is to relate the various explanatory filters as they have evolved.

I begin on the left side of figure 8.1, with Dembski's (1998c) questions: "(1) Does a law explain it? (2) Does chance explain it? (3) Does design explain it?" He continues: "I argue that the Explanatory Filter is a reliable criterion for detecting design. Alternatively, I argue that the Explanatory Filter successfully avoids false positives. Thus whenever the Explanatory Filter attributes design, it does so correctly" (107).

The central portion of figure 8.1 reflects the EF of Dembski (1999), where it was presented as a conventional flowchart. Here we note two important refinements: the notions of complexity and specification. The probabilistic nature of Dembski's argument is fully realized in *No Free Lunch* (2002b), represented on the right side of figure 8.1.

To implement the EF, we begin with the observation of an event. If no natural law–like explanation for the event is possible, Dembski asks whether there is a chance explanation. If no chance explanation is possible, then

Dembski decides that the event is the result of design. Design in the model is the default result when natural and chance explanations fail.

The notion of specified complexity provides a more-positive criterion. Dembski draws on Behe (1996) and his notion of irreducible complexity by equating specified complexity with irreducible complexity as the signature of design. Dembski (1998a, 46–47) offers an example of the EF's design detection in which a teacher receives two nearly identical student papers. He proposes two hypotheses: independent authorship and plagiarism. Dembski assigns independent authorship as the chance hypothesis and plagiarism as the design hypothesis, even though both outcomes are the result of intelligent action.

This assignment raises two significant points. First, Dembski admits that context determines how these hypotheses are to be classified as chance or design, leaving significant ambiguity in the classification. Second, his example fails to consider relevant alternate hypotheses: the students collaborated, the papers were nearly identical because of limited school resources, both students plagiarized a third party, or both were independently assisted by a third party (such as a tutor) and had no knowledge of the other's paper. These possibilities are all distinct from Dembski's plagiarism (design) hypothesis contrasted with his independent authorship (chance) hypothesis, which presumes that the students randomly generated identical papers and that the exclusive alternative is design.

Considering Dembski's plagiarism example, Fitelson et al. (1999) observe:

> It is important to recognize that the Explanatory Filter is enormously ambitious. You don't just reject a given Regularity hypothesis: you reject all possible Regularity explanations (Dembski 1998:53). And the same goes for Chance—you reject the whole category: the Filter "sweeps the field clear" of all specific Chance hypotheses (Dembski 1998:14, 52–53). We doubt that there is any general inferential procedure that can do what Dembski thinks the Filter accomplishes. (3)

They further point out: "Suppose you have in mind just one specific regularity hypothesis that is a candidate for explaining [event] E: you think that if E has a regularity-style explanation, this has got to be it. If E is a rare event, the filter says to conclude that E is not due to Regularity. This can happen even if the specific hypothesis, when conjoined with initial condition statements, predicts E with perfect precision" (3). The periodic observations of comets are a powerful example of rare natural events due to necessity, which, before Edmond Halley's research, were widely considered supernatural phenomena.

Without complete knowledge of all possible hypotheses, we cannot correctly assign chance and design hypotheses within the explanatory filter. It is entirely unclear how or even whether Dembski's explanatory filter could deal with multiple hypotheses, although Elliott Sober (in press) presents a likelihood method for detecting design that could offer some help to the EF. It is trivial to propose situations in which applying the EF serially to all possible hypotheses would require infinite time.

The Explanatory Filter and the Special Sciences

Dembski characterizes his method of design inference as equivalent to procedures used by archaeologists (when they recognize an artifact) and forensic scientists (when they assess a death scene). He observes that detecting design is basic to many human enterprises, listing copyright and patent offices, as well as cryptographers and detectives. In a key paper influencing both Dembski and the larger ID movement, Walter L. Bradley and Charles Thaxton (1994, 198–201) include an extended discussion of the analogical method in scientific reasoning. Their thesis, based on that of Thaxton et al. (1984), is that the origin of life is too improbable to be accounted for by any scientific explanation, so there must have been a creator. They cite archaeology, forensics, and the search for extraterrestrial intelligence (SETI) as ordinary scientific endeavors that detect (or search for) intelligent action, and they claim to apply the same reasoning to argue for a creator of life. Dembski has taken up this claim as a mantra:

> Within biology, Intelligent Design is a theory of biological origins and development. Its fundamental claim is that intelligent causes are necessary to explain the complex, information-rich structures of biology, and that these causes are *empirically detectable*. To say intelligent causes are empirically detectable is to say there exist well-defined methods that, on the basis of observational features of the world, are capable of reliably distinguishing intelligent causes from undirected natural causes. Many special sciences have already developed such methods for drawing this distinction—notably forensic science, cryptography, archeology, and the search for extraterrestrial intelligence (as in the movie *Contact*). (Dembski 1998c, 16–17; see also Dembski 1998f, 2000, 2001a, 2002c)

Recently, he has progressed from the claim that intelligent-design creationism is similar to archaeology, forensics, and SETI to the claim that ID actually subsumes them:

The fundamental idea that animates intelligent design is that events, objects, and structures in the world can exhibit features that reliably signal the effects of intelligence. Disciplines as diverse as animal learning and behavior, forensics, archeology, cryptography, and the search for extraterrestrial intelligence thus all fall within intelligent design. Intelligent design becomes controversial when methods developed in special sciences (like forensics and archeology) for sifting the effects of intelligence from natural causes get applied to natural systems where no reified, evolved, or embodied intelligence is likely to have been involved. (Dembski 2001a, n.p.).

The Explanatory Filter and Archaeology

Archaeologists spend a considerable part of their professional lives identifying and interpreting artifacts and groups or associations of artifacts. Associations of artifacts, such as a hearth or a trash pit, are called features when they represent a discrete behavior or activity. When the associations consist of multiple activities or behavior clusters within a limited geographical area, we call them sites. We also recognize that some aspects of intelligence are expressed in the artifacts and the associations we study. Superficially, archaeology and intelligent design might seem to be easily fused. There are, however, three distinctions that exclude the intelligent-design argument from the sort of archaeological association that Dembski assumes.

Archaeologists know precisely the identity of our designers, their fundamental needs, their available materials, and their range of means to manipulate those materials. Our close kin and we ourselves are the designers. Physics, chemistry, geology, and engineering provide our knowledge of their materials and means. Bradley and Thaxton (1994) cited archaeology merely as an example of analogical reasoning; they showed better understanding than Dembski.

Archaeologists have excavated sites that are the production of extant non-human intelligences: chimpanzees (Max Planck Research 2002, Mercader et al. 2002). Christophe Boesch and Hedwige Boesch (1983) are among the early researchers who have contributed to the extensive documentation of chimpanzee culture—that is, the geographically delimited sets of behaviors and technologies that differentiate different populations within the same subspecies (Chimpanzee Cultures 2003).

Archaeologists have excavated the remains and tools of our nonhuman ancestors and cousins as well. How are these discoveries recognized as artifacts? We have three sources of information: practical experience with the materials used, evaluation of objects in their context, and the commonality between contemporary behaviors and ancient behaviors.

Rather than William Paley's (1802) famous watch on the ground, consider the stone hammer as our diagnostic artifact. Archaeologists recognize stone hammers as built objects and in fact distinguish several different classes of hammer. Do we apply the explanatory filter? No, we do not. That would require us first to exhaust all possible explanations based on law or regularity, then all chance explanations, and then finally arrive at the notion that the object was designed.

Instead, we use what Dembski calls side information: independent knowledge about how different kinds of rocks behave when banged together. At some time, a stone is transformed into an artifact with a single blow from another stone. The stone is now a hammer. It is virtually impossible, however, to recognize this transformation if the action of stone striking stone is not repeated. The diagnostic features of stone hammers are actually the result of repetitive use, each incident of use causing the object to become more easily recognized.

When evaluating the possibility that a rock is a hammer stone, archaeologists do not want a high rate of false negatives (rejecting many slightly used hammers) nor a high rate of false positives (misidentifying plain rocks as tools). But for most purposes, a small number of false-positive errors is preferable to a massive rate of false-negative errors. Dembski aims to get no false-positive errors, so the explanatory filter must reject all slightly used stone hammers or else allow a flood of false positives—that is, classifying all stones of appropriate size and material as artifacts. Dembski is forced to accept a rate of false negatives unacceptable to archaeological research. In contrast, chimpanzee technology can be recognized and studied by the application of criteria identical to that used on human remains.

Hammer stones can also be compared to spider webs or beaver dams, which are results of simple operations repeated over time. According to Dembski (2000), beaver dams are the products of intelligence: "Consider beaver dams. They are not the product of human intelligence nor are they the product of Darwinian causes, but we are not ignorant of their causes. Beaver intelligence is responsible for beaver dams. (Note that invoking the Darwinian mechanism to explain why beavers build dams is not illuminating because if beavers didn't build dams, the Darwinian mechanism would readily account for this as well.)" (23).

There are two errors in this statement. One is a simple error of fact: beavers do not always build dams. They respond to specific environmental clues by piling up sticks and mud and do not build dams when they live on the shallow margin of a lake or large river, where these clues are absent. The other error is that, contra Dembski, this fact is obvious from a Darwinian perspective.

This reveals a problem in Dembski's criteria for intelligence. To be able to place a stone hammer in the "intelligently designed" bin, where we know it belongs, the explanatory filter will not only join human and chimpanzee intelligences but also the spider web, the beehive, and the beaver dam. Few would find this classification credible, and even then they would likely reject the notion of invertebrate intelligence (Cziko 2000). This drives the explanatory filter deep into the realm of false-positive assignments, which Dembski acknowledges are fatal to his scheme.

For us to properly reject spider webs as intelligently designed objects, we must again invoke more and more side information. In fact, the important work is all taking place on the side, not in the explanatory filter. Dembski fails to offer a usable definition of intelligence that can differentiate the instinctual behavior of spiders from that of humans. This fact is related to the student plagiarism example and is reason enough for archaeologists to ignore the EF.

The second difficulty is that, unlike ID, archaeology draws upon a vast literature of direct observational studies (ethnography) and an established base of replications (experimental archaeology). It relies tremendously on direct observation of behavior. Stone arrowheads were recognized to be human products only after the New World lithic (stone) technologies were documented, even though their existence was commonly known before then (Grinsell 1976).

This documentation was important in establishing that humans had existed contemporaneously with extinct European megafauna. Before this direct evidence, many rational people seriously thought these stone projectiles were "elf darts" used by fairies or witches to kill cattle (Hood 2003). Those supernatural beliefs were countered by materialists, who argued that the ancient artifacts were the natural products of lightning.

The realization that elf darts and thunder stones were ancient stone artifacts produced a new problem. Objects similar to stone tools were discovered in European gravel beds that held no other indication of human presence. These eoliths (dawn stones) were broken pieces of flint or chert with some properties in common with confidently designated artifacts. But the lack of any other human association was disquieting. Ultimately, they were rejected as human products because of differences between the types of fracture lines, surface conditions, and gross shape (Barnes 1939).

Since the 1950s, archaeologists have increasingly relied on replication studies. For example, they may manufacture stone tools, which they then use to butcher a carcass. Such studies allow us not only to observe the finished product of human building but also to recognize that the residues of these activities are as revealing as any other class of object. Archaeologists can in this manner gain interpretive advantage in the unintended consequences of tool building—a sort of "unintelligent design." Intelligent-design proponents can never offer such empirical basis because they refuse to explicitly identify the designer they advocate.

The third distinguishing characteristic between ID and archaeology is that, when diagnosing an artifact or an association, we do not rely on exhaustive exclusion of necessity before recognizing human behavior, nor do we invoke nonmaterial entities acting by unknown means as the default process for producing an object. Indeed, the EF is typically the opposite of how an archaeologist approaches an object. We first try matching the unknown object to known artifact classes. Once again, we rely on what Dembski regards as side information. We next consider how the object may have been modified. For example, the presence of incisions or grinding marks may indicate a built object. We then consider the associations of the object under examination: is it associated with an archaeological site?

In this last sense, consider an unidentified object found beneath a fallen wall. Because all the associations of the object are artifactual, the object is at least associated with human activity. Archaeologists even have a category of objects called manuports: totally unmodified natural objects found in unexpected locations, such as a hammer stone before it has ever been used or discarded food residue such as shells or bones.

Consider an example from the site-level class of association: a faunal midden. A midden is narrowly defined as deposited kitchen wastes. In practice, it is the hard-tissue remains of fauna and chemical alteration of the surrounding soil. Middens most commonly encountered in the coastal regions where I work contain shells of marine invertebrate species from several different environments (tidal mudflats, rocky shorelines, and sandy surf zones), bone from fish, and terrestrial species in a nonmarine soil that is a very dark color. We can have confidence in the identification of a midden only if the mollusks whose remains we examine have habits that cannot miraculously change overnight. We can rely on soil geology and recognize that pelagic fish cannot migrate over land, and we do not invoke unspecifiable properties of a miraculous flood to explain away stratigraphic associations.

There are several problems with Dembski's ambition to subsume archaeology as an aspect of his design theory. They reduce to this: archaeologists do not assume that natural law fails to explain a given phenomenon. They assume, rather, that natural law indeed explains it.

The Explanatory Filter and Forensics

Dembski (1999, 141) likens the explanatory filter's work in detecting design to a forensic scientist's work in detecting murder. Suicides, accidents, and murders, however, are not causes of death. These are socially constructed categories that are used to sort out kinds of deaths, not how death occurred. These categories have been subject to redefinition since the first such recorded categories in the Code of Hammurabi. Blood loss, shock, and tissue damage due to gunshot are causes of death that could be attributed to suicide, murder, accident, or justified homicide. Dembski repeatedly confuses these categories, as did Bradley and Thaxton (1994). Forensic scientists do not investigate murders per se. Rather, in the relevant context, we investigate death.

Forensics is largely an applied science. There are as many approaches to criminal investigation as there are criminals and investigators. Over the past century, however, an increasing amount of actualistic research (research conducted in natural settings, with very few controlled variables) has guided the examination and interpretation of death scenes. The portion of this research that is limited to field examination has been surveyed in the recent Federal Bureau of Investigation bibliography compiled by Michael J. Hochrein (2003). At 490 pages, this work is a powerful contrast to the missing original research of intelligent design.

Forensic science is like all other historical sciences in that practitioners rely on analogy, direct observation, replication, and the applications of basic sciences. The University of Tennessee Anthropology Research Facility (the Body Farm) is world-famous for actualistic research on the decomposition of human cadavers in natural settings. Studies on blood-spatter patterns, how to use a backhoe on a grave, and fly-maggot growth influenced not only by climate but even by a body's drug content can all be found in Agent Hochrein's bibliography.

The work of a forensic scientist begins with basic questions: are the remains human? How has the body been manipulated post mortem? What are the demographic characteristics of the dead person(s)? And then, what were the physical means by which death occurred? These are all naturalistic, physical mechanisms. If the cause of death cannot be determined, there is no further forensic science progress in the case. Period.

Consider the example I described at the beginning of this chapter: the woman whose leg fed some coyote puppies. Necessity has not explained her death (we don't know how she died); chance does not explain her death (we don't know how she died, but human death under circumstances that lead to predator gnawing is ipso facto complex). Strictly following the eliminative guide of the explanatory filter forces us to conclude design but can provide no help as to further action. Again, the real work is done with the side information, not the explanatory filter.

Even with the cause of death identified, the manner of death is neither obvious nor absolute. This fact is even clearer when we consider different kinds of homicides. Historically, what is considered murder is highly variable. Homicide interpretation is a kind of narrative construction. Suicide satisfies Dembski's requirements for a complex death that is the product of a purposeful intelligence while presenting fewer ambiguities than death at the hands of another.

Consider a death scene:

> The deceased was a young man in his early twenties. The body was discovered in the kitchen of his home 24 hours after he failed to arrive for work. There was considerable alcohol in his stomach, and a blood-alcohol test revealed that he was intoxicated at the time of death. The deceased was found suspended from a rope attached to a sturdy metal ring mounted in a roof joist. The rope was tied at the ring with bowline knot and was attached around the deceased's neck by a noose. There was a thin coating of oils and dust on the ring and nearby ceiling and no wood, paint, or plaster debris on the floor, indicating that the ring had not been recently installed. The film of oils and dust was disturbed or missing at the contact of the rope and the metal ring, which, together with the lack of oils or dust on the rope's upper surface, where it was tied to the metal ring, suggests that the rope was not regularly attached to the ring. A 24-inch-tall stool was on its side approximately 4 feet from the deceased's feet. There was a phone book (approximately 3 inches thick) on the floor below the deceased. The phone book cover had a tear and an impression similar to the feet of the stool. The impression was discernable for approximately 20 pages of the phone book. The deceased was naked. There was semen on the floor, which DNA analysis identified as the deceased's. Medical examination indicated that death was caused by strangulation.

How should we recognize this death? What does the explanatory filter tell us? Suicide by hanging is less common than gunshot or drug overdose, but neither is it uncommon. There is no evidence of a second party, and the knots are well tied. These are consistent with suicide. We also saw that that the ceiling ring was not a recent installation and that the page impressions and the tear in the telephone book indicate that considerable force was placed upon it. The weight of the deceased on the stool balanced with one leg on the telephone book would produce similar impressions. This evidence is inconsistent with suicide.

In what sequence do we take up the possible death characterizations presented by this case? Clearly, the dead man was alone at the time of death, and there is no evidence that he was forced by any other person to commit the acts that led to his death. So we can exclude murder, as I promised earlier. A forensic scientist would obviously exclude natural causes, not from exhaustive elimination but from experience. That exclusion leaves us with either chance (accident) or design (suicide) as the possible categorizations of this young man's death. As in the plagiarism example, both hypotheses represent events that were the result of intelligent action.

It may seem obvious that the death is suicide, but is it? The explanatory filter cannot hope to resolve the question. We need to consider the victim's motivation and intent, a consideration expressly denied by intelligent-design theorists. The death conforms to known sexual-behavior patterns, and forensic investigators can therefore recognize that accidental death is the most likely categorization in the case just presented.

Let us consider some additional examples. Followers of the recent Christian faith tradition of snake handling interpret Bible verses Mark 16: 18–20 and Luke 10: 19 to promise that the faithful will have supernatural immunity from venomous snakes. As a sign that the God of scripture is in the world, believers hold snakes and, in some congregations, ingest poison as part of their worship practice (Kimbrough 2002, anonymous 2003b).

On 24 July 1955, George W. Hensley, the founder of the American Holiness snake-handling movement, was bitten for the uncounted and last time. He refused medical treatment, as he had on many prior occasions. The following morning, at the age of 75, Hensley was dead of snakebite, just as his wife had died before him. Some of his followers believe that he died of a stroke.

On 3 October 1998, John Wayne "Punkin" Brown, Jr., was preaching at the Rock House Holiness Church in northeastern Alabama. With him was his 3-foot-long timber rattlesnake. The snake bit Brown on his finger, and the 34-year-old collapsed and died within 10 minutes. He had survived an estimated 22 snakebites. Brown's family does not rule out the possibility that his death was due to a heart attack rather than the snakebite.

In 1991 Glendel Buford Summerford, pastor of the Church of Jesus with Signs Following, was convicted of trying to kill his wife with poisonous snakes. Summerford, a snake-handling preacher, forced his wife at gunpoint to place her arm into a box full of rattlesnakes. The court found him guilty of attempted murder (Covington 1996).

In 1941, the state of Georgia outlawed snake handling in religious services. Georgia's statute was the most severe of several southern states and included the provision: "In the event, however, that death is caused to a person on account of the violation of this Act by some other person, the prisoner shall be sentenced to death, unless the jury trying the case shall recommend mercy." The reasoning behind a capital felony charge was that, if someone violated the snake-handling law and a death occurred, then the death was a murder. Georgia later repealed the law (Burton 1993, 81).

These snakebite deaths are alternately found in courts to be suicide, homicide, and accident. As Ted Olsen (1998) observed about Hensley's death, "Officials, showing a complete misunderstanding of Hensley's faith, listed his death as suicide" (n.p.). I argue, however, that Hensley's and Brown's deaths were accidental. Each knew that he was taking risks, but they had both successfully met these risks before. To the believer, death from these bites is tied directly to God's rejection of the deceased, which makes their surviving followers' suggestions that the deaths were due to stroke or heart attack understandable. I also agree with the jury's finding in the Summerford case.

The key point is that the explanatory filter and the entire ID rubric cannot distinguish whether these events were suicide, murder, accident, or divine retribution. Dembski cannot tell you what category they belong to based on his EF. The real world is a hard place to sort out.

Conclusion

In a recent essay, Robert Hedges (2003) remarked on the peculiar problems of understanding past behavior: "Archaeology suffers further difficulties in its reconstruction of the past, for here human behaviour is central and so we must engage, somehow, with the mental world of our forebears. As human mentality can encompass the most sophisticated acts (of deception, for example), it is not always satisfactory to rely on the present to explain the past, or to attempt to interpret behavioural evidence in a purely rational way" (667).

Dembski claims it is not necessary to have knowledge about a designer's nature or about the means that a designer used to impose its will on the material universe. Supporting this assertion are his reiterated versions of this statement: "There is a room at the Smithsonian filled with objects that are obviously designed but whose specific purpose anthropologists do not understand" (Dembski 1998g, n.p.; 1999, 157; 2002b, 147). His statement is untrue; there is no such room (Shallit 2002a).

Archaeologists nevertheless routinely recognize as artifacts objects that have no known purpose and whose functions we are unlikely ever to know. But in every instance we recognize them by the simple observation of marks: the pits, scratches, polish, grinding, burning, fracture, and so on that are the unambiguous indication of manufacturing. Dembski (1999, 141) admits that any false positives generated by the explanatory filter destroy his theory. We have seen that the filter is capable of explaining very little; so simple an event as one rock hitting another leads to a cascade of false negatives and false positives. The false-positive problem is more severe in the forensic examples; in the cases considered, the EF would incorrectly return a design verdict: each of those deaths was the direct result of purposeful actions, yet death, let alone suicide, was never intended.

Paley's famous watch may still lie on the ground, and it is still the inspiration of the intelligent-design creationist. But why should we think that Paley was correct that a naïve observer would assume the watch to be an artifact? Paley knew what a watch was and at least in general outline knew how one was formed. Can someone without any knowledge or even awareness of metallurgy, gears, or springs correctly discern the nature of a watch? Would that person necessarily recognize it to be a built object and reject a supernatural origin?

As with Paley's watch, all the serious work of detecting design in Dembski's scheme is done on the side, where we must create specifications to fit the events. Dembski will protest that side information is epistemically independent of events, that he does not draw targets around spent arrows. This protest is falsified if he insists that archaeology and forensics are subdisciplines of intelligent design. Dembski's explanatory filter can be made congruent with these sciences only if all the serious work is moved away from the filter and onto the side information.

Natural theology presumed that the universe was the product of God's will and that the rational study of the universe would reveal God's will on a par with the direct revelation of holy scripture. Two hundred years later, the intelligent-design movement is a desperate attempt to find in the universe unambiguous evidence that God exists at all.

Dembski raises the bar higher than he can jump and then ducks underneath it.

Chapter 9

Playing Games with Probability

Dembski's Complex Specified Information

JEFFREY SHALLIT AND WESLEY ELSBERRY

THROUGHOUT SCIENCE, random chance is an accepted component of our explanation of observed physical events. In chemistry, the ideal-gas laws can be explained as the average behavior of the random motion of molecules. In physics, the concept of half-life tells us what percentage of radioactive nuclei can be expected to decay within a given time period, even if we cannot identify, before the fact, which ones specifically will survive. In biology, random mutations and genetic drift are two of the probabilistic components of the modern theory of evolution.

But chance cannot explain everything. If we were to draw letters at random from a bag of Scrabble tiles, and the resulting sequence formed the message CREATIONISM IS UTTER BUNK, we would be very surprised (notwithstanding the perspicacity of the sentiment). So under what circumstances can we reject chance as the explanation for an observed physical event?

Let's be more specific. Suppose I entered a room alone, shut the door, and flipped a coin 50 times, recording the outcomes as heads (H) or tails (T). I then came out of the room and showed you the record of coin flips. Imagine that I produced this list of outcomes:

A: HHHTTHTHTHTTHHTHTHTHTTHTTHTHTHHHHHTHTHTHTHHT
TTHHHTHTHHHT

No one would be surprised in the least. But what if I produced this record?

B: HHHHHHHHHHHHHHHHHHHHHHHHHHHHHHHTTTTTTTTTTTTTTT TTTTTTTTTTTT

You would probably view with skepticism my claim of having produced it through random coin flips, and you would seek an explanation other than random chance. Perhaps I was really using two coins, one of which had two heads, the other two tails, and I accidentally switched between the two halfway through. Or perhaps I simply made up the record without flipping a coin at all. Can your skepticism be given a rigorous theoretical basis?

It does no good at all to say that B is a very unlikely outcome. According to standard probability theory, A is just as unlikely as B. In fact (assuming a fair coin), events A and B both occur with this probability,

$$\frac{1}{2} \times \frac{1}{2} \times \ldots \times \frac{1}{2} = 2^{-50} \qquad (1)$$

In other words, the probability is about 10^{-15}, or about 1 in a million billion. Yet B seems to us a much more unlikely result than A. We have stumbled on what appears to be a paradox of probability theory.

It is not a new paradox. James Boswell (1740–95), the biographer of lexicographer and essayist Samuel Johnson, wrote this about the events of 24 June 1784:

> I recollect nothing that passed this day, except Johnson's quickness, who, when Dr. Beattie observed, as something remarkable which had happened to him, that he had chanced to see both No. 1 and No. 1000, of the hackney-coaches, the first and the last; "Why, Sir, (said Johnson,) there is an equal chance for one's seeing those two numbers as any other two." He was clearly right; yet the seeing of the two extremes, each of which is in some degree more conspicuous than the rest, could not but strike one in a stronger manner than the sight of any other two numbers. (Boswell 1983, 1319–20)

French mathematician Pierre Simon Laplace (1749–1827) discussed the paradox in his 1819 *Essai philosophique sur les probabilités* :

> On a table we see letters arranged in this order, C o n s t a n t i n o p l e, and we judge that this arrangement is not the result of chance, not because it is less possible than the others, for if this word were not employed in any language we should not suspect it came from any particular cause, but this word being in use among us, it is incompara-

bly more probable that some person has thus arranged the aforesaid letters than that this arrangement is due to chance.

This is the place to define the word *extraordinary*. We arrange in our thought all possible events in various classes; and we regard as *extraordinary* those classes which include a very small number. Thus at the play of heads and tails the occurrence of heads a hundred successive times appears to us extraordinary because of the almost infinite number of combinations which may occur in a hundred throws; and if we divide the combinations into regular series containing an order easy to comprehend, and into irregular series, the latter are incomparably more numerous. (Laplace 1951, 231)

These remarks of Boswell and Laplace suggest a possible resolution of our paradox. In flipping a fair coin 50 times, some outcomes fit a short, simple pattern,

B: HHHHHHHHHHHHHHHHHHHHHHHHHHHHTTTTTTTTTTTTTTT TTTTTTTTTTTT

whereas others do not:

A: HHHTTHTHTTHHTHTHTTHTTHTHHHHHTHTHTHTHHT TTHHHTHTHHHT

The number of very simple patterns is small, so when a record that fits such a pattern is produced, we might legitimately reject "flips of a fair coin" as a valid explanation.

So far we have spoken imprecisely. What, exactly, is a valid pattern? How many valid patterns are there, and what does it mean to say this quantity is "small"? We will take up these questions later in the chapter. But now it is time to see how our paradox and its resolution can be misused to reach extraordinary conclusions.

Let's start with the Bible codes.

Bible Codes

In 1994, three Israelis, Doron Witztum, Eliyahu Rips, and Yoav Rosenberg, published a controversial paper in the journal *Statistical Science*. They claimed to find patterns that could not be explained by chance in the Hebrew text of the biblical book of Genesis (Witztum et al. 1994). Their unstated implication was that finding these patterns was so improbable that it suggested a divine origin for Genesis. Michael Drosnin (1998) then took this thesis to extremes, claiming to find biblical codes predicting the assassination of Yitzhak Rabin and retrodicting the Kennedy assassinations, the Oklahoma City bombing, and the election of Bill Clinton.

The patterns found by Witztum et al. (1994) were based on something they called an equidistant-letter sequence (ELS). An ELS in Genesis is a subsequence of the letters of the Hebrew text, in which letters are chosen according to an arithmetic progression: first, we examine the letter at position x; then the letter at position $x+a$; then $x+2a$; then $x+3a$; and so on. Witztum et al. claimed to find in Genesis equidistant-letter sequences giving the names of famous rabbis close to their birth or death dates and argued that the results were highly unlikely to have occurred by chance. The editor who published the original paper writes, "None [of the reviewers for or editors of *Statistical Science*] . . . was convinced that the authors had found something genuinely amazing. Instead, what remained intriguing was the difficulty of pinpointing the cause, presumed to be some flaw in their procedure, that produced such apparently remarkable findings. Thus, in introducing that paper, I wrote that it was offered to readers 'as a challenging puzzle'" (234).

The puzzle was later resolved by Brendan McKay, Dror Bar-Natan, Maya Bar-Hillel, and Gil Kalai (1999), who analyzed the results of Witztum et al. (1994) and found they could be easily explained by wiggle room in the data. For example, there was great flexibility in the choice of the particular rabbis searched for and in the choice of exactly how birth and death dates were represented. More important, the names of the rabbis themselves presented many choices: should we search for Maimonides, Rabbenu Moshe ben Maimon, or Rambam (all of which are legitimate forms of the name of this famous scholar)? Small changes in these choices substantially decreased the statistical significance of the results, so McKay et al. concluded that the data were tuned for the tests.

Thus, we see how the solution to our probability paradox can be misused. If we do not specify ahead of time precisely what patterns of observed events we regard as noteworthy, we run the risk of incorrectly rejecting chance as an explanation because we can construct a pattern to fit almost any outcome. Further, if an outcome's record is very long, we can pore over the symbols at length until we find something we regard as noteworthy. To illustrate this point, mathematician Brendan McKay (1997) found in the text of *Moby Dick* equidistant-letter sequences that "predict" the assassinations of Indira Gandhi; Martin Luther King, Jr.; and John F. Kennedy. He writes, "No laws of probability are violated here, or even stretched a little. . . . The reason it looks amazing is that the number of possible things to look for, and the number of places to look, is much greater than you imagine."

An ingenious way to avoid the phony-pattern problem was discovered by the Russian mathematician Andrei Kolmogorov (1965) and is often called algorithmic information theory. (Essentially the same ideas were explored independently, at about the same time, by Ray Solomonoff and Gregory Chaitin.) Kolmogorov actually had two bright ideas: first, to restrict the set of valid patterns to those checkable by a computer and, second, to assess a cost based on the complexity of the pattern. We now turn to his theory.

Algorithmic Information Theory

Kolmogorov complexity is the principal tool of algorithmic information theory. Roughly speaking, the Kolmogorov complexity of a string of bits x is the length of the shortest combination (P,i) of program and input that will produce x when the computer program P is run on the input i. (By the length of (P,i) we mean the number of bits used to write it down.) This complexity is denoted by $C(x)$ and is sometimes called the information contained in x. Note that the running time of the program P does not figure at all into our considerations here; P could produce x in 1 microsecond or 1 millennium, and $C(x)$ would be the same.

A string x has low Kolmogorov complexity if there is a short program P and a short input i such that P prints x when run on input i. For example, the bit string

11111111111111111111111111

has low Kolmogorov complexity because it can be generated by the program

print 1 n times

together with the input $n = 26$.

It appears we have defined $C(x)$ rigorously, but what programming language should we use? Unfortunately, there is no natural or universally agreed-upon choice. Should we use Java, C, APL, Pascal, FORTRAN, or something else entirely? In fact, mathematicians typically use none of these, preferring a programming model called the Turing machine (named after its inventor, Alan Turing; in this case the program P is actually an encoding of a Turing machine that can be interpreted by a "universal" Turing machine). Each choice of programming language might result in a different value of $C(x)$. But an important result called the invariance theorem states, roughly speaking, that the Kolmogorov complexity relative to one programming language L_1 is equal to the complexity relative to another language L_2, up to a fixed additive constant that depends only on L_1 and L_2. So it doesn't really matter what programming language we choose to express our program P in, as long as we make a single choice and stick to it.

Kolmogorov complexity is strongly related to optimal lossless data compression. Lossless data compression may be familiar as the technology that allows you to store a large file in an encoded form that (often) takes up less space on your hard drive, using a command such as Zip.

If (P,i) is the shortest program-input pair that produces x, we can think of (P,i) as the best possible way to compress x. If we wanted to store x, we could store (P,i) instead, since we could always recover x by running the program P on the input i. In the case of strings containing a small amount of information, such as $111 \ldots 1$, it evidently makes sense to store them in some compressed form rather than write out all those 1's.

Not every string, however, can be compressed. For each possible length, there is at least one string x that is not compressible at all. That is, there is at least one string x such that the compressed representation (P,i) has at least as many bits as x itself. Such strings are termed *random*. Note that this is a definition of the term *random*; a string that is random in the Kolmogorov sense possesses many of the properties we associate with being random.

Similarly, at least half of all strings of length n cannot be compressed by more than 1 bit, at least three-quarters cannot be compressed by more than 2 bits, and so on. It follows from this theorem that most strings have relatively high Kolmogorov complexity.

We have seen that $C(x)$ can be very small for highly patterned strings. Is there a limit on how big it can be? The answer is that we always have $C(x) \le |x| + c$ for some fixed constant c. Here $|x|$ is shorthand for the length of, or number of bits in, the string x. To see this fact, observe that every string can be compressed by outputting the program "print the input" together with the input x itself. Here c represents the length of the program "print the input."

It follows that, for long strings x, the quantity $C(x)/|x|$ is a number that is between 0 and a little more than 1 and measures the complexity of the string x. Table 9.1 illustrates how strings can be classified.

We are finally ready to understand Kolmogorov's solution to our probability paradox and the phony-pattern problem. The idea is that we allow as a legitimate pattern for a string x any combination of program and input (P,i) such that P produces x on input i. Some patterns are better than others because they are shorter; we can write down P and i with fewer bits. We assess a cost based on the length of (P,i), and the minimum possible cost is precisely the Kolmogorov complexity $C(x)$. Thus, no pattern is really phony; there are just short patterns and long ones, and we prefer, or are impressed by, the short ones.

Table 9.1.
Classification of strings by Kolmogorov complexity.

$C(x)/\lvert x \rvert$ close to 0	Highly compressible Ordered Fits a simple pattern Low information Nonrandom Infrequent A randomly chosen string has this property with low probability
$C(x)/\lvert x \rvert$ close to 1	Highly incompressible Fits no simple pattern High information Random Frequent A randomly chosen string has this property with high probability

Suppose we form a string x that is n bits long from the flips of a fair coin, where 0 represents heads and 1 represents tails. For t between 0 and n,

the probability that $C(x) \leq t$ is then $\leq 1/2^{n-t}$ (1)

If t is small and n is large, then this probability is very small. We have thus obtained a quantitative version of Laplace's observation: the shorter the pattern, the more unlikely the result! This marvelous solution has been known to mathematicians for quite a while (see, for example, Levin 1984, Gács 1986, Li and Vitányi 1988, Kirchherr et al. 1997).

That's the good news. Now the bad news: $C(x)$ is actually uncomputable. That means there is no computer program that will take arbitrary strings x as input and unerringly return their Kolmogorov complexity $C(x)$. So it seems that our whole solution has collapsed.

There is a way out, however. We can approximate $C(x)$ by producing some program-input pair (P,i) such that P outputs x on input i. Then the length of (P,i) overestimates $C(x)$.

To sum up, suppose we are given a string x that consists of n 0's and 1's. We then try to find a short program P and an input i such that P produces x on input I—the shorter the better. How we do so is immaterial; we could, for example, try to deduce P and i by inspection, or we could use a commercially available compression routine such as Unix's Compress. Suppose the length in bits of (P,i) is t. We can then say that the chance that a series of flips of a fair coin produces a string that can be compressed as well as x is $\leq 1/2^{n-t}$. If t is very small compared to n and x really represents the flips of a fair coin, we have been extraordinarily fortunate to witness such a rare event—so fortunate that we will probably seek an explanation for *x* other than a series of flips of a fair coin.

What other explanations are possible? There are many. The string *x* could indeed represent the flips of a coin, but a biased coin. For example, the coin could be weighted so that heads occurs, on average, 19 out of 20 times. In that case, a record of flips such as

C: 00000000000000000000000100000000000000000000

would not at all be surprising, despite the very low Kolmogorov complexity of the string. Remember: our analysis applies only to flips of a *fair* coin, where heads and tails occur with equal probability. If the probability *p* of getting a head is something other than $1/2$, we need a different formula to replace equation (1), a formula that depends on the entropy of the associated probability distribution.

Another possible explanation is that *x* does not represent the flips of a fair coin but the output of some simple computational process. For example, explaining a result like

B: HHHHHHHHHHHHHHHHHHHHHHHHHHHHHHTTTTTTTTTTTTTTT TTTTTTTTTTTT

is easy if we postulate that the record was not generated by flips of a fair coin but (say) by recording H at position *i* if the temperature in Tucson, Arizona, is above 20 degrees Celsius at 5*i* minutes past 5 P.M., and T otherwise. Finally, the string *x* could represent the flips of a fair coin, but by someone who is adept at cheating: at flipping the coin in such a way that the desired result nearly always occurs.

Can we decide among these various possibilities? Not on the basis of the record of events alone. We need additional evidence.

Dembski's Complex Specified Information

Finally, we come to the topic of this essay: the pseudomathematics of William Dembski. Dembski is a theologian, philosopher, and mathematician who claims that his mathematics proves that biological organisms were designed by an intelligent being. This design, he claims, cannot be accounted for by the generally accepted mechanisms of evolution, such as mutation, natural selection, and genetic drift.

In broad outlines, Dembski's claims are nothing new. William Paley (1802) argued 200 years ago that, if we find a watch, we can deduce the existence of a watchmaker because the watch has a purpose, and any small change in the size or placement of the parts would render it unusable for that purpose.

Despite this lack of novelty, Christian apologist William Lane Craig has called Dembski's work "groundbreaking" (Dembski 1999, opening blurb). Journalist Fred Heeren (2000) describes Dembski as "a leading thinker on applications of probability theory." Yet according to the American Mathematical Society's online version of *Mathematical Reviews*, a journal that attempts to review every noteworthy mathematical publication, Dembski has not published a single paper in any journal specializing in applied probability theory and only one peer-reviewed paper in any mathematics journal at all. On the back of Dembski's *Intelligent Design: The Bridge between Science and Theology* (1999), University of Texas philosophy professor Robert Koons calls Dembski the "Isaac Newton of information theory," an endorsement Koons (2001) repeated at a recent conference. But according to *Mathematical Reviews*, Dembski has not published any papers in any peer-reviewed journal devoted to information theory. Could the effusive praise of Heeren and Koons be unwarranted?

Dembski thinks that intelligence has a magical power that permits it to do something that would be impossible through natural causes alone. Furthermore, this power is detectable: "When intelligent agents act, they leave behind a characteristic trademark or signature" (Dembski 1999, 127). He calls this trademark of intelligence complex specified information or specified complexity.

Roughly speaking, Dembski's specified complexity is defined as follows: we witness a physical event E. We then somehow assign E to some class Ω of possible events. Next, we try to find a pattern T to which E conforms. If T is suitably independent of E (Dembski calls it epistemic independence), we say E is specified. We then compute the probability p that a randomly chosen outcome from Ω would match T. If this probability is less than or equal to 2^{-k}, we say that E has at least k bits of specified complexity.

An event E with enough bits of specified complexity possesses complex specified information (CSI). In this case, Dembski asserts that E arose by intelligent design, not random chance or natural causes (or a combination of those). How many bits are required? Dembski is inconsistent. Sometimes a very small number suffices: Dembski (1999, 159) claims that "the sixteen-digit number on your VISA card" or "even your phone number" contain CSI. In other cases, at least 500 bits are required.

In its rough outlines, Dembski's specified complexity strongly resembles the well-known Kolmogorov solution to the probability paradox we have already outlined. Unfortunately, Dembski's approach serves only to muddle the well-known solution and make it unworkable. As we will see, Dembski's specified complexity has none of the properties he claims it has.

Playing Games with Probability

As with the Bible codes, pseudomathematics often takes the form of bogus probability arguments. Pseudoscientists love probability because it offers a quick route to their desired conclusion. Want to prove evolution impossible? Just use some unjustified estimates about the probability of various events, and presto! You've proved what you want, and "mathematically" to boot. This leads us to the first problem with Dembski's reasoning: the hazy rationale for the assignment of probabilities to events.

What is probability, precisely? There are many different philosophical interpretations. A frequentist would say that probability deals with many repeated observations: the more events we observe, the more likely a measured probability will be close to the "true" probability. Consider flipping an ordinary pair of dice. The probability of obtaining the outcome 7 for an ideal pair of dice is 1/6. But due to imperfections in the dice and slight variations in the weights of the sides, the probability for any real pair of dice will not be 1/6 but some close approximation to it. What is that probability? We may be able to deduce it from a physical model of the dice. But we can also measure it empirically with high confidence, by flipping the dice millions of times.

On the other hand, the events Dembski is most interested in are singular: receipt of a message from extraterrestrials, the origin of life, the origin of the flagellum of the bacterium E. coli, and so forth. By their very nature, such events do not consist of repeated observations; hence, we cannot assign to them an empirically measured probability. Similarly, because their origins are obscure and we do not currently have a detailed physical model, we cannot assign a probability based on that model. Any probability argument for such events therefore affords a splendid opportunity for mischief.

Dembski himself is inconsistent in his method of assigning probabilities. If a human being was involved in the event's production, Dembski typically estimates its probability relative to a uniform probability hypothesis; let's call this the uniform-probability interpretation. For Dembski, a Shakespearean sonnet exhibits CSI because it would be unlikely to be produced by choosing several hundred letters uniformly at random from the alphabet.

On the other hand, if no human being was involved, Dembski nearly always bases his probability calculations on the known causal history of the event in question; let's call this the historical interpretation. This flexibility in the choice of a distribution allows Dembski to conclude or reject design almost at whim.

Sometimes he uses these two different methods of calculating probability in the same example. Consider his analysis of a version of Richard Dawkins's (1986, 46–48) METHINKS IT IS LIKE A WEASEL program. In this program, Dawkins shows how a simple computer simulation of mutation and natural selection can, starting with an initially random 28-letter sequence of capital letters and spaces, quickly converge on a target sentence taken from *Hamlet*. In *No Free Lunch*, Dembski (2002b) writes,

> Complexity and probability therefore vary inversely—the greater the complexity, the smaller the probability. It follows that Dawkins's evolutionary algorithm, by vastly increasing the probability of getting the target sequence, vastly decreases the complexity inherent in that sequence. As the sole possibility that Dawkins's evolutionary algorithm can attain, the target sequence in fact has minimal complexity (i.e., the probability is 1 and the complexity, as measured by the usual information measure is 0). Evolutionary algorithms are therefore incapable of generating true complexity. And since they cannot generate true complexity, they cannot generate true specified complexity either. (183)

Here Dembski seems to be arguing that we should take into account how the phrase METHINKS IT IS LIKE A WEASEL is generated when computing its complexity or the amount of information it contains. Since the program that generates the phrase does so with probability 1 and $2^{-0} = 1$, the specified complexity of the phrase is 0 bits.

But in other passages of *No Free Lunch*, Dembski seems to abandon this viewpoint. Writing about another variant of Dawkins's program, he says,

> the phase space consists of all sequences 28 characters in length comprising upper case Roman letters and spaces. . . . A uniform probability on this space assigns equal probability to each of these sequences—the probability value is approximately 1 in 10^{40} and signals a highly improbable state of affairs. It is this improbability that corresponds to the complexity of the target sequence and which by its explicit identification specifies the sequence and thus renders it an instance of specified complexity. (188–89)

Here his use of a uniform probability model is explicit. Later, he says,

> It would seem, then, that E has generated specified complexity after all. To be sure, not in the sense of generating a target sequence that is inherently improbable for the algorithm (as with Dawkins's original example, the evolutionary algorithm here converges to the target sequence with probability 1). Nonetheless, with respect to the original uniform probability on the phase space, which assigned to each sequence a probability of around 1 in 10^{40}, E appears to have

done just that, to wit, generate a highly improbable specified event, or what we are calling specified complexity. (194)

In the latter two quotations, Dembski seems to be arguing that the causal history that produced the phrase METHINKS IT IS LIKE A WEASEL should be ignored; instead, we should compute the information contained in the result based on a uniform distribution on all strings of length 28 over an alphabet of size 27. (Note that 27^{28} is about 1.2×10^{40}.) The uniform-probability interpretation and the historical interpretation can give wildly differing results, and Dembski apparently cannot commit himself to one or the other, even in the context of a single example.

The second problem with Dembski's work concerns selecting the reference class of events to which an observed event E belongs. Observed physical events do not typically come with probability spaces attached. If we encounter a string of a thousand 0's, should we regard it as a string chosen from an alphabet consisting of just the single 0 or the alphabet $\{0, 1\}$? Should we regard it as chosen from the space of all strings of length 1000, or all strings of length ≤ 1000? Dembski's advice is unhelpful here; he says the choice of distribution depends on our "context of inquiry" and suggests "erring on the side of abundance in assigning possibilities to a reference class" (Dembski 2002b, sec. 3.3). But following this advice means we are susceptible to dramatic inflation of our estimate of the information contained in a target, because we may well be overestimating the number of possibilities. Such an overestimate results in a smaller probability, and the smaller the probability, the larger the number of bits of specified complexity Dembski says the event contains.

Specification

The third problem with Dembski's work is his recipe for determining the pattern T to which an event E conforms. According to Dembski, not every pattern is permissible; he spends a lot of time discussing how to separate legitimate patterns (specifications) from phony ones (fabrications). Among other things, his framework demands that a valid pattern must be "explicitly and univocally" identifiable from the background knowledge of an intelligent agent (Dembski 2002b, 63).

It follows that CSI is not a fixed mathematical quantity but can vary according to the observer. For a French speaker, the sentence HONI SOIT QUI MAL Y PENSE is immediately identifiable from background knowledge, but a non-Francophone will find it meaningless gibberish. This relativity of measurement is a very strange property of Dembski's information: the value of mathematical quantities does not usually depend on who computes them.

Indeed, specification has much in common with other forms of pseudoscience, such as René Blondlot's N-rays (Gratzer 2000). Blondlot was a French physicist who claimed to have identified a new form of radiation with remarkable properties. Oddly enough, N-rays could not be detected by everyone; one skeptic was told dismissively, "Your eyes are insufficiently sensitive to appreciate the phenomena." Eventually, N-rays were discovered to be nothing more than a figment of Blondlot's imagination. In the same way, Dembski's evanescent notion of specification seems to depend on the cultural sensitivity of the intelligent agent involved.

Let's return to the pattern T. Exactly how is it determined? Dembski is inconsistent here. Sometimes he claims it doesn't matter how: in fact, we are allowed to simply "read [T] off the event E" (Dembski 1998a, 146). But other times he says that T must be determined "without recourse to the actual event" E (Dembski 2002b, 18).

This latter requirement is very strange. How could anyone verify that an event E actually conforms to a pattern without recourse to E—that is, without actually examining every bit of the event in question? To illustrate this, consider Dembski's discussion of the bit string shown in the first example in figure 9.1, which is a variation on a signal received by fictional researchers in the movie *Contact*. As Dembski describes it, t consists of blocks of consecutive 1's separated by 0's; the lengths of these blocks encode the prime numbers from 2 to 89, with extra 1's at the end to make the length exactly 1000. Dembski suggests the specified complexity of this sequence implies design.

Dembski says t is specified. Let us now restate his specification as S = "a string containing the unary representations of the first 24 prime numbers, in increasing order, separated by 0's, and followed by enough 1's at the end to make the string of length 1000." (The unary representation of a number n is just a string of n 1's.) Presumably Dembski believes it self-evident that S could enable us to identify t "without recourse to the actual event."

But we cannot; in fact, S is *not* a specification of the actual printed sequence! A careful inspection of the string presented on pages 143–44 of *No Free Lunch* (Dembski 2002b) reveals that it inexplicably omits the prime number 59. In other words, the string Dembski actually presents is the string shown in the second example in figure 9.1. So in fact, our proposed specification S does *not* entail t' but t.

The point is not that Dembski made a silly mistake but that the pretense that we could ever come up with a pattern for a string without actually looking at the string is nonsensical. Dembski's epistemic-independence criterion is therefore invalidated.

There is another problem with Dembski's notion of specification. Because

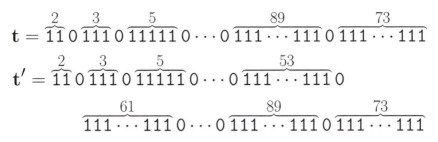

Figure 9.1. Upper, a string of bits encoding the prime numbers from 2 to 89, as described by Dembski. The last 73 1's are included to make the total number of bits supposedly 1000. Lower, the string of 1000 bits actually presented by Dembski. The encoding of the prime number 59 is missing from his string. After Dembski (2002b, pp. 143–44).

it is so vague and because he does not assess a cost based on the length of the specification, it is susceptible to a problem known as the heap paradox (Sainsbury 1995). The heap paradox asks, "When does a collection of grains of sand become a heap?" Clearly one grain of sand does not form a heap. And if n grains of sand do not form a heap, then adding one grain to get $n+1$ grains is not going to suddenly change a non-heap to a heap. We conclude that no heaps of sand exist. Yet they do.

This paradox can be resolved by realizing that the term *heap* is not a rigorous, black-and-white classification. There are degrees of heapness. A pile of ten grains of sand is only very slightly a heap, while a pile of a million grains is very strongly a heap. Once we measure heapness numerically, the paradox disappears.

In the same way, Dembski insists that specification is a black-and-white classification: an event is either specified or it isn't. But it doesn't make any sense to say, for example, that the text of Shakespeare's *Hamlet* is specified, but exactly the same text with an extra comma at the end is not. If x, a string of 0's and 1's, has a specification T, then the string $x0$ containing an extra 0 at the end can be specified simply by amending T to say "and add another 0 at the end." We can continue this process ad nauseum; without assessing a cost, *every* event is specified.

Thus, Dembski's notion of specification discards the crucial ingredient: a cost assessed on the length of the pattern's description. If we charge a cost based on the length of the pattern's description, then we get essentially the well-known Kolmogorov solution to the probability paradox we have already discussed. In this case, the specified complexity of a binary string x turns out to be essentially $|x| - C(x)$, where $|x|$ denotes the length of the string x and $C(x)$ is the Kolmogorov complexity. We might call this quantity "specified anti-information" because it is close to the negation of what mathematicians and computer scientists usually mean by information—namely, $C(x)$.

Finally, we point out one more inconsistency in Dembski's treatment of specification. According to its formal definition, a specification is supposed to consist of a lot of mathematical apparatus: a space of events, a rejection region, a rejection function, and certain real numbers. But when it comes to applying the definition of specification, Dembski doesn't use his own framework. Consider his discussion of the specification of the flagellum of *Escherichia coli*: "in the case of the bacterial flagellum, humans developed outboard rotary motors well before they figured out that the flagellum was such a machine" (Dembski 2002b, 289). What, precisely, is the space of events here? What are the rejection function and the rejection region? Dembski does not supply them. Instead he says, "At any rate, no biologist I know questions whether the functional systems that arise in biology are specified" (289). That may be, but the question is not "Are such systems specified?" but "Are the systems specified in the precise technical sense that Dembski requires?" This is equivocation at its finest (or worst). The bottom line is that Dembski's notion of specification is so muddled that we cannot say with certainty whether a given outcome is specified or not. Like Blondlot's N-rays, the existence of CSI seems clear only to its discoverer.

The Law of Conservation of Information

Physics, as we learned it in high school, is bristling with laws: Newton's inverse-square law, Boyle's ideal gas law, the second law of thermodynamics, and so forth. A common misconception about physical laws is that they are analogous to human-made laws. (As the joke goes, "Speed limit 186,000 miles per second: not just a good idea, it's the law!") But they aren't. In physics, the term *law* is just shorthand for a simple mathematical relationship between physical quantities—a relationship that has been confirmed over and over again in thousands of experiments. True, physical laws don't always apply under all conditions: for example, the ideal-gas laws are just approximations and fail at low temperature and high pressure. Nevertheless, law in physics means something that is an accurate description of phenomena over a wide range of conditions.

Calling something a law has good public relations value. (Zellers, a Canadian department store chain, claims that, in their stores, "the lowest price is the law.") There is a certain cachet about the word *law*, which makes anything labeled as a law seem inviolate. If you have a new theory and call it a law, who can argue with it? (Even better, capitalize it and call it a Law.)

And so we come to Dembski's most grandiose claim, his law of conservation of information (LCI). It has, he tells us, "profound implications for science" (Dembski 2002b, 163). One version of LCI states that CSI cannot be generated by natural causes; another states that neither functions nor random chance can generate CSI. We will see that there is simply no reason to accept Dembski's "law" and that his justification is fatally flawed in several respects.

Suppose we have a space of possible events, each with an associated probability. To keep things simple, let's suppose our class is the set of all strings of 0's and 1's of length n, where each symbol occurs with probability $1/2$, so each possible string has an associated probability of 2^{-n}. Now suppose we have a function f that acts on elements of this space, producing new binary strings of the same length. We write $f(x) = y$, meaning that f acts on a string x, producing a string y.

Dembski argues that, no matter what the function f is, if the string $y = f(x)$ has a certain amount of specified complexity, then x has at least the same amount. His argument uses basic probability theory and is technical, so we won't repeat it here. But it is flawed, and the flaw depends on Dembski's ambiguous notion of specification. Since y has specified complexity, there is an accompanying specification T. Dembski now claims that $f^{-1}(T)$, the inverse function of f applied to T, is a specification for x. But remember that patterns are supposed to be "explicitly and univocally" identifiable with the background knowledge of an intelligent agent. Why should $f^{-1}(T)$ be so identifiable? After all, the claim is supposed to apply to all functions f, not just the f known to the intelligent agent A computing the specified complexity. The function f might be totally unknown to A—for example, if f occurred in the long-distant past. In fact, there is no reason to believe that A will be able to deduce that x is specified, so x has no specified complexity at all for A. Thus, applying functions f can, in fact, generate specified complexity.

To look at a more-concrete example, let's suppose y is a string of bits containing an English message, perhaps in ASCII code, and f is an obscure decryption function, such as RSA decoding. We start with a string of bits x having apparently no pattern at all; we apply f, and we get y, which encodes the message CREATIONISM IS UTTER BUNK. Any intelligent agent can recognize y as fitting a pattern (for example, the set of all true English sentences), but who will recognize x as fitting a pattern? Only those who know f. This objection alone should be enough to convince the reader that the law of conservation of information is bogus. But there is yet another problem.

The second problem with LCI is that, as we have observed, Dembski uses two different methods to assess the probability of an observed outcome: the uniform-probability interpretation and the historical interpretation. Dembski's mathematical argument applies only to the historical approach. But as we have seen, this implies that the complete causal history of an observed event must be known; if a single step is omitted, we may estimate the probability of the event improperly. This is fatal to Dembski's program of estimating the specified complexity of biological organisms, because the individual steps of their precise evolutionary history are largely lost with the passage of time.

Although his mathematical justification for LCI depends on the historical interpretation, he rarely appeals to it. Instead, he uses the less-demanding uniform-probability interpretation. But LCI fails for this interpretation.

Let's look at an example. Consider again the case in which we are examining binary strings of length n and define our function f to take a string x as an input and duplicate it, resulting in the string $y = xx$ of length $2n$. For example, $f(0100) = 01000100$. Under the uniform probability interpretation, when we witness an occurrence of y, we will naturally view it as living in the space of strings of length $2n$, with each such string having probability 2^{-2n}. Furthermore, any y produced in this manner is described by the specification "the first and last halves are the same." A randomly chosen string of length $2n$ matches this specification with probability 2^{-n}. Thus, under the uniform probability interpretation, we have two alternatives: either every string of length n is specified (so specification is a vacuous concept) or f actually produces specified complexity!

The law of conservation of information is no law at all.

The Growth of Complexity

Dembski's specified complexity is an important piece of a broader creationist argument: that the complexity of biological organisms is so large that they could not have arisen through natural processes; hence, intelligent intervention is needed to explain them. We have now seen where this argument breaks down. Specified complexity is not a well-defined quantity, and there is no law of conservation of information that ensures that complexity cannot grow over time.

Where, then, does the complexity of living things come from? We cannot answer this question until we have a satisfactory definition of the word *complexity*. If complexity means Kolmogorov complexity, then there is nothing to explain since random processes can generate as much complexity as we like. If complexity means specified anti-information as discussed previously, then as much complexity as we like can be generated by simple computational processes.

How do simple computational processes arise in nature? This is an interesting question whose answer we are just beginning to understand. It turns out there are many naturally occurring tools available to build simple computational processes. To mention just four, consider the recent work on quantum computation (Hirvensalo 2001), DNA computation (Kari 1997), chemical computing (Kuhnert et al. 1989, Steinbock et al. 1995, Rambidi and Yakovenchuk 2001), and molecular self-assembly (Rothemund and Winfree 2000). Furthermore, it is now known that even very simple computational models, such as Conway's game of Life (Berlekamp et al. 2003), Langton's ant (Gajardo et al. 2002), and sand piles (Goles and Margenstern 1996) are universal and hence compute anything that is computable. Finally, in the cellular automaton model, relatively simple replicators are possible (Byl 1989).

Dembski's claim that CSI is a trademark of intelligent agents is therefore suspect. CSI provides no way to separate the actions of intelligent agents from the results of simple, naturally occurring computational processes. Indeed, intelligent agents themselves could well be naturally occurring computational processes.

Conclusion

The bottom line is that Dembski's specified complexity or complex specified information is an incoherent concept. It is unworkable, is not well-defined, and does not have the properties he claims for it. Even Dembski himself, in attempting to calculate the specified complexity of various events, uses an inconsistent methodology. Most important, specified complexity does not provide a way to distinguish designed objects from undesigned objects.

Biochemist Russell F. Doolittle (1983) once remarked, "The next time you hear creationists railing about the 'impossibility' of making a particular protein, whether hemoglobin or ribonuclease or cytochrome-c, you can smile wryly and know that they are nowhere near a consideration of the real issues" (261). That same wry smile might be useful to keep handy when reading Dembski's claims.

Chapter 10

Chance and Necessity— and Intelligent Design?

TANER EDIS

PHYSICAL SCIENTISTS are used to a bottom-up view of the world. At the most basic level, elementary particles and interactions are described by mathematically challenging theories. These fundamental theories, though difficult to grasp, nevertheless reveal a remarkably simple structure in our world.

Then things get complicated. Elementary particles bind to make nuclei and atoms. Atoms form molecules. Studying molecules, we start to go beyond physics. Even so, our fundamental theories appear to be capable of accounting for all that is built upon them, including the intricacies of chemistry. In fact, this bottom-up approach has worked so well that we take it for granted.

When confronted with a puzzle at the fundamental level, we might find that something entirely new is afoot, as happened when we had to invoke a new force to explain atomic nuclei. But when we are confronted with a puzzle at a higher level, we do not suspect something entirely new. On the contrary, we assume that the higher levels must be reducible to the lower. No one thinks, for example, that there is a life force associated with organic molecules. No physicist postulates a new fundamental interaction to explain a phenomenon such as superconductivity, no matter how puzzling. So now our question becomes how to explain the complexities of our world, starting from a simple realm of fundamental particles and forces.

To connect what we see with more-basic levels of physics, we can try to show that a complex phenomenon is due to physical law, or necessity. Water, for example, has its properties because of the way in which electromagnetic forces bring oxygen and hydrogen together. Alternatively, we may find

randomness, or chance. There is no way to predict individual decays of ura-
nium nuclei; these are random events.

In fact, physics relies exclusively on combinations of chance and necessity. A snowflake has a hexagonal pattern due to physical laws. But the form
of an individual snowflake is also due to chance. The exact pattern realized
depends on the environmental noise that shapes the freezing snowflake in arbitrary ways. Indeed, our everyday world appears to have been "frozen out"
from among the endless possibilities compatible with basic physics through
just such a process of symmetry breaking. Our physical world is a result of
chance and necessity; moreover, chance and necessity are inseparable, just as
the randomness of a coin toss is inseparable from the symmetry between heads
and tails (Edis 2002, 103–7).

Today, many of our sciences are concerned with complexity. And there
is much to study. In the world of basic physics, we do not find life, intelligence, or even our everyday sense of cause and effect. So we develop theories
of thermodynamics, of computation, of evolution, and more—all attempting
to explain complex aspects of our world. Interestingly, none of these theories
is very sensitive to the details of the underlying microscopic physics; Darwinian evolution, for example, can work no matter what the physical source of
random variations. And all these theories rely on chance and necessity to generate complexity.

Can Physics Do It All?

A natural question is whether chance and necessity can bear such a burden.
Physicists may be able to explain how macroscopic physics such as the fluid-
ity of water or the nature of temperature emerges from the microscopic level.
Physical science, however, is about mindless phenomena explainable by chance
and necessity. Why should we think this sort of explanation works across the
board and indeed applies to our own minds?

The intelligent-design (ID) movement denies that chance and necessity
are sufficient, insisting that intelligent action is a fundamental principle beyond basic physics. ID targets biological evolution, in part because living things
appear to have a special kind of order. Chance and necessity might assemble
a snowflake, but no matter what variation comes up, it is still a snowflake. Randomly rearranging a genetic code, in contrast, is extremely unlikely to give
us a viable animal. Only a tiny subset of all possible arrangements works. Pick-
ing out the functional code, when nonsense seems to be chemically just as
feasible, seems to call for intelligence. Moreover, many of the adaptive fea-

tures of living things are amazing in their ingenuity. Even if life started with simpler forms, some of it has increased in complexity in a way that demands genuine creative novelty. How can blind chance and rigid necessity give us creativity?

ID also gains plausibility from our common intuitions about minds—about intelligence itself. Again, it appears incredible that mere chance and necessity could give rise to intelligence; common sense suggests that intelligence must be a separate principle in the world. And if so, the clearest signature of intelligent action in the world would be the complex, purposeful structures due to intelligent design.

So the ID movement can be expected to criticize artificial-intelligence (AI) research, which suggests that machines, driven by chance and necessity, may be capable of exhibiting creative intelligence. Indeed, some ID proponents pronounce this aspect of AI fundamentally misguided (Dembski 1999, 216–18). Since AI today is more a promise of progress than a body of established results, biological evolution remains the more urgent concern for ID proponents. Still, because AI also threatens to reduce creativity to chance and necessity, an AI perspective should help illuminate the claims of ID.

Detecting Design, Denying Darwin

William Dembski, the leading theorist of the ID movement, tries to prove that chance and necessity cannot be creative. His approach relies on concepts of information and complexity similar to those used by computer scientists (see chapter 9 in this book). If correct, his work would also mean that building a true artificial intelligence is impossible.

Dembski starts by identifying a special kind of order shared by forms of life and artifacts of human design, capturing the notion that there is something to life that is different from the intricacy of a snowflake and similar to what we see in items designed for a function. In Dembski's terminology, artifacts and living things exhibit specified complexity. Then he argues that chance and necessity, including Darwinian variation and selection, cannot create the information inherent in specified complexity. So at some stage, an outside intelligence must have intervened (Dembski 1998b, 1999).

Imagine finding a piece of paper with letters printed on it. If these letters spell out a clear message, we think it is a product of intelligence. Even when we cannot figure out the precise meaning—if, for example, the message is written in a foreign language—we recognize features of human languages that make us think it is meaningful, and we again attribute it to intelligence.

If, however, we were to find a haphazard jumble of letters on the paper, something like IHJFL/BLACV?GYUFHRFWWVHBMD . . . , we would think differently; apparently, a mindless process had produced meaningless nonsense. We would find necessity in the fact that only ordinary characters appeared and chance in the haphazard nature of the result, but we would not find intelligent design.

Dembski tries to formalize such intuitions, stating that meaningful messages exhibit contingency, complexity, and specification. Contingency expresses the requirement that many different messages, even complete gibberish, can be written on a piece of paper with equal ease. Heavy objects around us all fall downward. But this fact does not signify design, since heavy objects fall by necessity. By contrast, letters on a piece of paper are under no such physical constraint.

If we found a piece of paper covered with alternating letters, HTHTHT HTHTHTHTHTHTHT . . . , we need not suspect an intelligence behind it. This pattern is not complex enough to be a message; a very simple computer program or just a faulty printer could have produced it. Chance and necessity will occasionally produce something meaningful—blind luck might result in a short statement like A CAT—but a complex piece of writing is extremely improbable.

More important, the properties of a real message can be specified before the fact without prior knowledge of what actually is on the paper. Otherwise, we may be tempted to think that a piece of apparent gibberish is a coded message. Any sequence of letters whatsoever can be "decoded" to mean absolutely anything, if we devise a suitably convoluted decoding procedure after the fact.

Dembski (1998a) claims that we can test for contingency, complexity, and specification to filter out what is due to chance and necessity; what is left must be a result of design. Pure chance will almost certainly give us gibberish, which may look complicated but does not exhibit specified complexity. Necessity can constrain a system to make it unsuitable for conveying information; at best, it can produce only the simplest of sequences.

Dembski's argument, translated into AI terms, is that, by mindlessly manipulating symbols, a machine can play with form but can never generate content. A computer is all syntax and no semantics. Chance and necessity can never produce meaning. It is no surprise that such claims sound familiar; this is a very common conceptual objection to AI. Dembski just adds the more-concrete claim that specified complexity is just what an intelligence can produce and a machine cannot.

Now, Dembski's way of inferring design has not impressed workers in relevant fields such as information theory, complexity, or probability theory. As things stand, his work appears to suffer from numerous technical problems, especially with the details of specification (Fitelson et al. 1999). The kindest assessment would be that Dembski has not yet provided us with a rigorous way to detect design.

Still, it would be premature to dismiss the design inference. A better way to think of Dembski's work might be as an attempt to formally capture the special kind of order common to living creatures and many of the products of our own minds. There is indeed something special about life, and we would not be able to do science without informal ways of identifying special patterns that demand explanation. It seems worthwhile to try to formalize such intuitions, even if early tries are likely to be too simplistic. So ID proponents might argue that Dembski's work, even with all its technical deficiencies, is a starting point for further research.

Let us assume that we have a workable procedure to detect and describe the special kind of order in life and designed artifacts and that this procedure is an improved version of Dembski's proposal. What, then, are the prospects for ID?

Even with an improved procedure, serious problems would remain. Chiefly, just identifying a special order need not have any implications for evolution. Biologists, after all, adopted evolution to explain biological complexity without denying that this order was similar in some ways to what we see in artifacts. Evolution and intelligent design could both create complex order, but after obtaining detailed evidence about the history of life, biologists concluded that living things had evolved.

In fact, although Dembski tries to eliminate chance and necessity, his filter is apparently not fine enough to rule out Darwinian means (Perakh 2004, Elsberry 1999). He does not adequately address *combinations* of chance and necessity. The contingency, complexity, and specification test strains even to exclude something like a snowflake. A snowflake exhibits contingency, and since any individual configuration is very improbable, it is complex in Dembski's sense of the word. Someone who lacked our background knowledge of physics and chemistry could easily think of the hexagonal symmetry of a snowflake as a specification of an order inexplicable by pure chance or by necessity, thus inferring intelligent design.

No doubt the snowflake problem can be fixed by making the specification aspect of Dembski's scheme more sophisticated and removing the ambiguity in attributing a pattern to necessity or to design. Even so, building a snowflake, for all its intricacy, does not call for a very elaborate combination of chance and necessity. It is nothing like the incremental and continually branching process of Darwinian evolution. Since evolution builds on what comes before, it incorporates memory, producing a history. In Dembski's test for design, there is no history. Everything is static, flattened out, inferred from the end product alone.

This brings up a curious problem with his design inference. Lacking a sense of history, Dembski's test says nothing about when and how specified complexity was infused into a structure. So an improved version of the test might work yet say very little of significance concerning evolution. For example, all of the information we see in genetic material might be due to the initial conditions of the universe (Edis 2001). In that case, evolution could easily have taken place; it would now just be the way in which the information embedded in the microscopic physics became apparent at the macroscopic, biological level.

In other words, Dembski's version of ID is fully compatible with evolution in the sense of common descent. It is even compatible with variation and selection as the driving forces behind evolution, provided that we no longer understand the variation as truly blind but set so as to ensure our particular evolutionary history. Such a view would be congenial to many liberal theologians, but it does not sit well with the conservative position that Dembski (1999) takes. It would also be one of those grandiose but inconsequential philosophical notions that are easily ignored in the practice of science.

This lack of history has other strange implications. Dembski wants to look at an end product and infer intelligence behind it. If we find a piece of paper with writing on it, we think it has an intelligent source. If we find that a computer printed it out, we then think the computer could not have produced the specified complexity; the information must have already been contained in its programming. But then we may wonder if the same applies to human artifacts. A Dembski-style design inference does not force us to say that a human is the ultimate source of information any more than is a computer. Dembski certainly thinks that we genuinely create information rather than just act out what was set in the initial conditions of the universe. But his procedure for detecting specified complexity tells us nothing about the immediate source of information; we could, so far as it is concerned, be automatons.

In explaining complexity, ID relies on an analogy between human designs and living organisms. Just as there is an intelligence that built a computer, there must be an intelligence that designed cockroaches and cabbages. But any argument for ID based on Dembski's procedure must remain incomplete because whatever reason we have for attributing real creativity to ourselves has nothing to do with specified complexity.

Where Does Information Come From?

We should not push this criticism too far; Dembski need not do all the work for ID. Still, we need to ask why we attribute intelligence to humans and not computers, and we need to explore how Dembski's design inference is related to questions about creativity.

Dembski captures fairly well the reasons why we do not think a computer is intelligent. Clearly, a computer's output expresses the information in its input and its programming; it cannot add new content. It may print out a Shakespearean sonnet, but the creativity was all Shakespeare's. AI efforts often try to circumvent this preprogrammed rigidity by having us imagine very long, extremely sophisticated programs continuously interacting with the outside environment. But even then, the output would be determined entirely by the input and the programming. No true novelty could emerge. A computer program can preserve information, or it can degrade it. It changes how information is represented, which may be very useful indeed; but it cannot create information.

Darwinian evolution is different because it incorporates an element of chance—something not determined by programming or input. But Dembski (2002b) argues that this fact does not change the situation. Consider an example used by Richard Dawkins (1986, 46–48) to illustrate the incremental workings of evolution. Starting with a haphazard jumble of letters like BNNWPDF5YHBNSSR U*S!AQWEFG7C, imagine that we replicate it but with some randomly varying letters in each position. Then we let the variant that is closest to METHINKS IT IS LIKE A WEASEL survive to reproduce again. A machine that implements this variation-and-selection procedure will converge on METHINKS IT IS LIKE A WEASEL (see chapter 9 in this book). As Dembski observes, however, all the information in this sentence is built into the selection criteria. Again, the creativity resides outside the machine.

Dembski expands upon this theme by using the no-free-lunch theorem (Wolpert and Macready 1997). He assumes that Darwinian evolution is analogous to solving a problem by using a genetic algorithm. Let us say we have to search for a solution among a vast set of possible solution attempts, each of which can be better or worse than neighboring attempts that differ in small ways. This task can be conceived of as finding the high point in a very complicated fitness landscape—like trying to find the highest peak in a mountainous region when we cannot look at the landscape as a whole but are aware only of our immediate surroundings and our altitude (see chapter 11 in this book).

Genetic algorithms set about climbing mountains in a way that mimics evolution. We start with a population of solution attempts. Say we have flocks of sheep dotting the landscape. We then allow sheep who live at a higher altitude to reproduce more than those that live lower; maybe the mountain air is invigorating. We also allow random variations, in that a newborn sheep might live in a nearby spot slightly higher or lower than its parent (a single parent: there is rarely sex in genetic algorithms).

If we then follow our population of sheep over many generations, we will find that the flocks creep upward, eventually exploring the summits of local mountains. Given long enough, they will even find the highest peak. Genetic algorithms apply this principle of variation and selection to all sorts of problems, almost never involving sheep.

Now, genetic algorithms can indeed come up with unexpected and useful results. But they work well only on certain fitness landscapes. In fact, the no-free-lunch theorem states that, averaged over all fitness landscapes, every search procedure, no matter how seemingly insane, performs the same as every other. Dembski interprets this result to mean that, even if variation and selection were the means to construct an instance of specified complexity, the action still was not genuinely creative because it was merely using the information inherent in the fitness landscape. According to Dembski, even in much more complicated situations than METHINKS IT IS LIKE A WEASEL, the Darwinian mechanism can reveal only the information preexisting in its selection criteria.

We think that humans, in contrast to machines, are truly creative. Human artifacts may be genuinely new rather than just a revelation of preprogrammed information. Our intelligence is apparent in our flexibility and our ability to make meaningful rather than haphazard choices. It appears that we are different from computers in that we can always do otherwise; we are not bound by preprogrammed rules. When confronted with a new problem, we may fail, but we are not guaranteed to fail. We always have the possibility of figuring out an ingenious new solution. Moreover, we learn and expand our capabilities. In fact, if anything is truly creative, truly intelligent, we must be it; humans are our defining examples here.

Contrasting intelligence with preprogrammed behavior is not just an ID preoccupation; it features in some long-standing lines of criticism of AI. The most interesting is the series of arguments based on Gödel's incompleteness theorem (Lucas 1961; Penrose 1989, 1994). Gödelian critiques emphasize that we can show that any computer program, no matter how elaborate, has blind spots. There is a point beyond which it must fail, where it cannot solve a new problem because it cannot figure out the appropriate new trick. And this happens solely because it is bound by rules; a human mathematician examining the program can always figure out precisely how to trip it up. In other words, computers are limited by their programs, lacking the creativity to go beyond them.

Dembski (1999, 219–22) acknowledges this critique, objecting only that it doesn't go far enough in drawing antimaterialistic conclusions. Additionally, he argues that including chance behavior does not change this picture. So let us summarize ID, Dembski-style: *no mechanism, relying on chance and necessity only, can generate specified complexity. Specified complexity can be detected in the products of intelligence; it is the signature of intelligent design. Humans can be truly creative—generate specified complexity—because they are not determined by preset rules or driven by chance.* Creative intelligence is supposed to produce outputs that no unaided mechanism can achieve.

Flexibility without Magic

There seems to be something compelling about such a claim; it is hard to deny its force. Nevertheless, it is wrong. We can say, very confidently, that chance and necessity can be genuinely creative. We can even say that in all likelihood our own creativity, our own intelligent designs, can be traced to chance and necessity. To see why, let us begin with the Gödelian objection to AI.

Computers are incapable of certain tasks: nonalgorithmic functions for which no finite program can be written. A classic example is Turing's halting problem. Computer programs can either halt, producing some sort of result, or get stuck and work forever to no end. It turns out that no possible computer program could scan other programs and always tell us whether they will halt (Boolos and Jeffrey 1989). This is a perfectly well defined task but beyond the reach of computers.

Unfortunately, we are no better at figuring out things like the halting function than our machines are. Such specific nonalgorithmic functions, called oracles by computer scientists, cannot be computed. But Gödelian arguments lead us to think that intelligence must be nonalgorithmic, not capturable by any finite set of rules. If that is so, then we must look for a kind of nonalgorithmicity that does not require specific oracles but ensures we are always flexible, not bound by rules.

If we allow a machine to use randomness—if we combine chance and necessity—we can get just this kind of flexibility (Edis 1998a). A string of random outcomes is the least rule-bound result possible. A random function is completely haphazard and thus entirely nonalgorithmic. In fact, it is useless except for one thing: giving us something completely novel, unconditioned by rules. So a touch of randomness might be just what is missing from an ordinary computer. The ability to flip a coin can make it a nonalgorithmic device, with flexibility to go beyond its initial program.

There are other reasons to suspect that randomness is the crucial ingredient. It has long been known that the ability to decide randomly is crucial for certain tasks. In game theory, for example, partly random behavior is often the best option against an opponent who can adapt to and exploit a regular strategy (Berger 1980, 10–13). In solving a predetermined problem, a machine that incorporates randomness is not more capable than an ordinary computer; it still cannot compute oracles. But in circumstances requiring a machine that does not get stuck in a rut of its own programming, randomness is just what is needed.

The flexibility conferred by randomness is still far from actual intelligence. But now we can use a completeness theorem (Edis 1998a), which demonstrates that the non–rule-bound flexibility characteristic of intelligence *must* be achievable through a combination of rules and randomness. The completeness theorem shows that any function whatsoever can be expressed as a combination of a random part and a finite algorithm or program part: chance and necessity. To construct an oracle, we would need to know the precise infinite random sequence defining that oracle. Knowing that sequence appears to be impossible; we have no reason to think that oracles actually exist or that anything is capable of the magical intuition required for oracular knowledge. There are, however, tasks that do not require a specific random sequence but for which any random function will do the job. These are tasks in which randomness serves as a source of novelty, preventing a machine from getting caught in ruts determined by its programming. Even the "program" such a machine executes can be subject to random variation.

Put another way, completeness means that tasks requiring specific oracles are the only ones that chance and necessity must necessarily fail at accomplishing. And no one can do these tasks anyway. In particular, specified complexity has nothing to do with oracles. If an improved version of Dembski's test one day reliably detects specified complexity in objects, that will be an interesting, perhaps even useful, development. But the complexity detected by this test will be accessible to mechanisms built out of chance and necessity.

So the attempt to go beyond chance and necessity by examining the complex products of intelligence cannot work. We still might be tempted to think that the algorithmic part of a machine's behavior must have an external source. But if so, this would mean that the reason we attribute intelligence to humans has nothing to do with the complex designs we perform or our flexible behavior. Not only is such a conclusion implausible, but it also undercuts the reason to look for an intelligence behind specified complexity in the first place.

Dembski's version of intelligent design is fundamentally mistaken. It does not fail just because of empirical inadequacies, although ID suffers plenty in that department. It suffers from more than technical deficiencies in methods of detecting design, although again there is no end of such problems. We can make an even stronger case for ruling out intelligent design as an independent principle, and this case is based on our knowing what chance and necessity can and cannot achieve. Specified complexity simply does not fall into the realm of what is beyond machines.

Darwinian Creativity

My claim about specified complexity needs to be fleshed out. Completeness is a very general result; it tells us nothing about *how* chance and necessity combine for the sort of creativity leading to organisms or artifacts, only that such creativity is possible. It tells us Dembski must be wrong, but it does not say where his mistake lies.

How can we get creativity out of the raw novelty that randomness provides? How can we incrementally build up complex machines? One way to do so has been known for some time: Darwin's mechanism. Biologists know, in considerable detail, how this mechanism works. Although formal demonstrations of Darwinian mechanisms gradually increasing information have so far been confined to simple simulations (Schneider 2000, 2002), there is no reason that such results cannot be generalized to more-complex and -realistic biological scenarios.

In fact, a striking development in recent decades has been the way in which Darwinian thinking has taken root outside biology. In particular, in AI and cognitive science, Darwinian approaches to the mind have become prominent. We have Gerald Edelman's (1992) neural Darwinism; the Darwin machines and variation and selection in the brain proposed by William Calvin (1996); memes, a nongenetic form of replicating information operating in the realm of culture (Blackmore 1999, Aunger 2002); Daniel Dennett's (1995) multiple levels of Darwinian mechanisms depending on processes competing to assemble our stream of consciousness; and more. Researchers in machine intelligence are increasingly relying on Darwinian mechanisms to introduce creativity into machines—even beginning to explore art and original engineering designs (Fogel 2000, Bentley 1999). AI is no longer an enterprise devoted to canned, preprogrammed strategies; it includes open-ended, evolutionary behavior.

This is not to say that human creativity is even close to being fully explained. Darwinian variation and selection are almost certainly vital parts of the picture; but other mechanisms, such as conceptual blending (Fauconnier and Turner 2002), are bound to be crucial for the rich creativity we find in ourselves. We are just beginning to find out. Even now, however, we can say with some confidence that human intelligence is not something separate from a world of chance and necessity.

ID proponents like to portray not just evolutionary biology but also AI and cognitive science as stagnant fields, unable to overcome deep, persistent problems. They argue that conventional research cannot overcome those problems because of these fields' commitment to inadequate theories such as Darwinian evolution. Admitting intelligent design as a separate principle, they say, will clear the way, leading to the required breakthroughs. The actual developments in these fields, however, are very different. We keep making progress in understanding not just biology but also human intelligence itself in terms of chance and necessity.

What, then, of no free lunch, of METHINKS IT IS LIKE A WEASEL, of all the ways in which Dembski argues that chance and necessity can do no more than shuffle existing information? Dembski's mistake is subtle but straightforward: he conceives of evolution as a way to search for a solution to a predetermined problem. It is nothing of the sort. Darwinian evolution is creative precisely because nothing is predetermined and everything may be randomly modified.

If evolution truly was a search for a high point on a fixed fitness landscape, in the manner of a genetic algorithm, Dembski's argument might be plausible. In that case, allowing a machine to make random decisions would not change what it is capable of solving and what it is not. Then the kinds of search procedures that would work well or not would depend on the fitness landscape, so we might be tempted to think a Darwinian mechanism introduces no genuine novelty. The problem is set; therefore, finding the solution becomes a matter of letting the information inherent in the problem bubble up to the surface.

As biologists point out (Orr 2002), evolution is not at all like a search on a fixed fitness landscape. Living populations are not searching for a solution to a preset problem. Their fitness landscape is continually changing, and this change is largely due to the other organisms that make up an important part of an organism's environment. Even an organism's own reproductive strategy alters the fitness landscape. All that is important is being able to reproduce, and what works best at any one moment is not likely to remain so forever, since competitors are themselves always changing.

The no-free-lunch theorem that Dembski relies on does not apply when the fitness landscape changes in a way that depends on the population (Wolpert and Macready 1997). Indeed, being able to randomly alter strategy is important since competitors may adapt to any set strategy and exploit it. And a prime way to generate increasing complexity in biology is to have evolutionary arms races (Dawkins 1986, 178). In an arms race, competing populations can climb smooth hills of fitness constructed by the competition itself, using variation and selection.

Physicists working to explain complexity also recognize the importance of this point. Old-fashioned creationists often challenge evolutionists to explain how, in a world tending to disorder because of the second law of thermodynamics (see chapter 7 in this book), biological order is supposed to increase without intelligent intervention. Cast in physical terms, Dembski's arguments about chance and necessity being able only to preserve or degrade specified complexity are a close cousin of the creationists' second-law argument.

Spontaneous ordering of the sort we see in evolution can take place in systems driven away from thermodynamic equilibrium. One example is the universe after the big bang. The expansion of space means that the maximum possible entropy of the universe increases faster than the actual entropy. This gap creates opportunities for order to form. Evolution, in fact, works for just this reason (Edis 1998b). As species diversify, the diversity actually realized increases more slowly than does the number of all possibilities (Brooks and Wiley 1988).

In other words, assembling complexity through chance and necessity depends on an expanding set of possibilities. It requires a changing world, one in which, by accident, history can take a genuinely new path to the exclusion of others. In contrast, if we have a set destination, history is merely about success or failure. If all we had was a search for the best spot on a fixed fitness landscape, evolution could not even take hold, let alone be genuinely creative.

Chance and Necessity All Around

Darwin brought the study of life in line with physical science. In modern biology, life does not require special principles such as a life force; neither does it evolve through anything other than chance and necessity. Many have perceived this as a loss, as another triumph of the ugly urge to reduce everything to physics. Scientifically, however, joining biology and physics improved both disciplines. Biology became connected to the accomplishments of the physical sciences; it acquired an impressive unifying theory that made it more than a collection of unrelated items of information about life. Physicists in turn tested and expanded their own concepts in attempting to understand complexity. By learning how they were continuous with one another, both biology and physics became richer.

Something similar is happening today. We have just started to do real science about questions concerning human and machine intelligence. But already we have forged close links among physics, biology, and newer disciplines such as cognitive science and artificial intelligence. By sticking to chance and necessity, we have begun to appreciate the richness of complexity, of intelligence. Declaring these to be unassailable mysteries could never lead to such understanding. For mystery is all that the ID movement offers: a mysterious principle of intelligent action removed from any taint of mechanism.

Any prospect for such mystery, though, is fading. Evolution was first proposed as an alternative to intelligent design. Biologists did not directly challenge the notion that human artifacts were products of a mysterious principle; instead, evolutionists pointed out how living things differed from artifacts, looked at evidence about the history of life, and concluded that the diversity of life was best explained by a nonmysterious alternative.

Back then, we still might have thought that intelligence was something beyond chance and necessity. Today, it is becoming clearer how our own intelligence is rooted in mindless mechanisms. Intelligent design is not a separate principle; it is built up from blind chance and rigid necessity. So we can look back, and as ID advocates ask us, once again highlight the similarities between artifacts and forms of life. We can again entertain the notion that intelligent designs and life share a special kind of order, because design is just another manifestation of chance and necessity. Darwinian mechanisms can act directly to shape life, or they can work indirectly, through the variation and selection in our brains, which lets us design things.

Down deep, there are only chance and necessity.

Chapter 11

There Is a Free Lunch after All

William Dembski's Wrong Answers
to Irrelevant Questions

MARK PERAKH

The subtitle, "Why Specified Complexity Cannot Be Purchased without Intelligence," of William Dembski's (2002b) book *No Free Lunch* indicates that he perceives the no-free-lunch (NFL) theorems (Wolpert and Macready 1997) as pivotal to his thesis that "specified complexity cannot be purchased without intelligence." Indeed, many statements in Dembski's book emphasize the crucial role of the NFL theorems. In his response to a review rebutting his use of the NFL theorems, however, Dembski (2002a) claims that the theorems are secondary to his thesis, while his principal argument is related to a so-called displacement problem. In this chapter I show that neither the NFL theorems nor the notions related to the displacement problem support Dembski's thesis.

Critiques of Dembski (2002b) can be found in a number of publications (Wein 2002a, 2002b; Shallit 2002b; Rosenhouse 2002; Perakh 2001b, 2002a, 2002c, 2004a; Orr 2002; Van Till 2002). Here I will discuss only chapter 4 of that book, "Evolutionary Algorithms," concentrating on Dembski's use of the NFL theorems and on his discussion of the displacement problem.

Methinks It Is Like a Weasel—Again

In many of his publications, including *No Free Lunch*, Dembski repeatedly discusses Richard Dawkins's METHINKS IT IS LIKE A WEASEL evolutionary algorithm, trying to prove its fallaciousness. This is how Dawkins (1986) describes the weasel algorithm: "It . . . begins by choosing a random sequence of 28 letters. . . . It now 'breeds from' this random phrase. It duplicates it repeatedly, but with a certain chance of a random error—'mutation'—in the copying. The computer examines the mutant nonsense phrase, the 'progeny' of the original phrase, and chooses the one which, however slightly, most resembles the target phrase" (47–48).

Dembski (2002b) sees an inadequacy in Dawkins's algorithm: it converges on a target phrase. He says, "choosing a prespecified target sequence as Dawkins does here is deeply teleological. . . . This is a problem because evolutionary algorithms are supposed to be capable of solving complex problems without invoking teleology" (182). But later he says, "An evolutionary algorithm is supposed to find a target within phase space" (203). Searching for a target *is* teleological.

Such inconsistency is Dembski's trademark. In any case, neither of his statements is correct. Evolutionary algorithms may be either targeted or targetless. Biological evolution, however, has no long-term target. Evolution is not directed toward any specific organism. The evolution of a species may continue indefinitely so long as the environment exerts selection pressure on that species. If a population does not show long-term change, it is not because that population has reached a target but because the environment, which coevolves with the species, acquires properties that eliminate its evolutionary pressure on the species. Dawkins's weasel algorithm, on the other hand, stops when the target phrase has been reached.

Dawkins (1986, 50) was himself the first to point out that his algorithm differs from biological evolution in that it proceeds toward a target. But then a model is not supposed to be a replica of the entire modeled object or phenomenon (Perakh 2002d); models replicate only those features of the modeled objects that are crucial for analyzing a specific, usually limited, aspect of the modeled object or phenomenon and ignore all the aspects and properties that are of minor importance. Dawkins's algorithm was designed to show that a combination of random variations with a suitable law can accelerate evolution by many orders of magnitude; the law in this case is selection. The algorithm indeed shows such an acceleration. As Dembski points out, a random search would require, on average, 10^{40} iterations of the search procedure. Dawkins's algorithm performs the task in about only forty iterations.

Dawkins's procedure is not a proof of evolution, but it is a valid demonstration of a very significant acceleration of evolution if a suitable law works along with random variations. That is why, as Dembski (2002b) laments, "Darwinists and even some non-Darwinists are quite taken" with Dawkins's example (183). It is, indeed, a good example.

Is Specified Complexity Smuggled into Evolutionary Algorithms?

Dembski suggests a modification of Dawkins's weasel algorithm. In his adjusted procedure, the algorithm will "pick a position at random in the sequence. . . . Then randomly alter the character in that position. If the new sequence has a higher fitness function than the old, keep it and discard the old. Otherwise keep the old. Repeat the process" (193). I will discuss the fitness function later in this chapter. For now, suffice it to say that the fitness function is the number of letters in the intermediate phrases that coincide with the letters in the same positions in the target phrase.

I see no substantial difference between the procedure described on pages 47–48 in Dawkins's book and that suggested by Dembski. In Dembski's view, while Dawkins's algorithm compares consecutive phrases with a target, his own modified algorithm "searches for the target solely on the basis of the phase space and the fitness function," hence "not smuggling in any obvious teleology" (194). But the only difference between Dawkins's algorithm and Dembski's modification is in the way in which they simulate mutations. Otherwise, both compare intermediate phrases with the target. In Dembski's version, the values of the fitness function are simply the counts of those letters in the intermediate phrases that coincide with the letters occupying the same positions in the target phrase. Indeed, we read, "As before, fitness is determined by how close a sequence is to the target sequence" (194). Therefore, Dembski's modified algorithm is as teleological as Dawkins's original algorithm.

Continuing, Dembski insists that evolutionary algorithms cannot generate specified complexity (SC) but can only "smuggle" it from a "higher order phase space" (194–96). This claim is irrelevant to biological evolution. In the case of the weasel algorithm, the outcome is deliberately designed. SC is injected into the algorithm through the fitness function. But since biological evolution has no long-term target, it requires no injection of SC. Natural selection is unaware of its result—the increased chance for having progeny. The advantage in proliferation occurs automatically. If Dembski thinks otherwise, he needs to offer evidence that extraneous information must be injected into the natural selection algorithm, apart from that supplied by the fitness functions that arise naturally in the biosphere. He provides no such evidence.

Furthermore, in Dawkins's weasel example, the evolutionary algorithm converges on a meaningful phrase: a quotation from Shakespeare. According to Dembski, the target phrase possesses SC. Michael Behe, in a foreword to Dembski (1999), gives an example. While the meaningful sequence METHINKSITISLIKEAWEASEL is both complex and specified, a sequence NDEIRUABFDMOJHRINKE of the same length, which is gibberish, is complex but not specified. Many of Dembski's statements scattered throughout his publications make it clear that Behe has indeed correctly reflected his position (Perakh 2001b, 2003), which is that a meaningless sequence possesses no SC.

On the other hand, Dembski (2002b, 195) indicates that Dawkins's algorithm could also be applied if the target phrase were gibberish. But if the target sequence is meaningless, then, according to Behe's quotation, it possesses no SC. If the target phrase possesses no SC, then obviously no SC had to be smuggled into the algorithm. Hence, if we follow Dembski's ideas consistently, we have to conclude that the same algorithm smuggles SC if the target is meaningful but does not smuggle it if the target is gibberish. This notion is preposterous because algorithms are indifferent to the distinction between meaningful and gibberish targets.

This inconsistency in Dembski and Behe's approach stems from the fact that the very concept of SC is contradictory. In fact, contrary to their notions, both a meaningful phrase and a string of gibberish are specified if the concept of specification is given back its commonsense meaning by clearing it of the embellishments and unnecessary complications suggested by Dembski (1998a, 1999, 2002d). By having written down a gibberish sequence, Behe has clearly specified it. As soon as it has been written, it becomes unequivocally distinguishable from any other sequence, which means that it is specified. Whether it is meaningful or gibberish is of no consequence (Perakh 2001b, 2004a).

Targetless Evolutionary Algorithms

While Dembski devotes much attention to Dawkins's weasel algorithm, he ignores another procedure designed by Dawkins: the biomorphs algorithm (BA). The BA differs from the weasel algorithm in that it has no target and is designed to illustrate a different aspect of evolution. The BA demonstrates how evolution, starting with a very simple progenitor, can generate unlimited complexity without a preselected target. Since its purpose is different, the features of evolution retained in this model are also different from the weasel algorithm.

The biomorphs algorithm has its own limitations. Dawkins (1986) points out that it "shows us the power of cumulative selection to generate an almost endless variety of quasi-biological form, but it uses artificial selection, not natural selection. The human eye does the selecting" (60–74). Thus, like every model, the BA is not a replica of reality but is adequate to illustrate an important feature of reality: the generation of complexity. Dembski's failure to discuss the targetless biomorphs algorithm undermines his critique of the weasel algorithm for its teleological features.

He does, however, discuss another model suggested and used by Thomas Schneider (2000, 2001a, 2001b), who claims that it is targetless. Schneider maintains that his algorithm generates biologically meaningful information from scratch—that is, without an input from intelligence. Dembski (2002b) disagrees: "The No Free Lunch theorems . . . tell us this is not possible" (215).

Proper analysis of Schneider's evolutionary algorithm would require a much more complex discourse than Dawkins's weasel algorithm, and I will not attempt to do so in this chapter. I will, however, address Dembski's main argument against Schneider's algorithm: whether it indeed generates information from scratch, or whether information is supplied by a hidden target of Schneider's program. Dembski's argument, based on the NFL theorems, misinterprets these theorems. He states repeatedly throughout his book that the NFL theorems prohibit generation of information without intelligence. In fact, they do nothing of the sort.

The No-Free-Lunch Theorems

What do the NFL theorems say about biological evolution or about evolutionary algorithms such as those developed by Dawkins (1986), Schneider (2000, 2001a, 2001b), Altshuler and Linden (1999), Chellapilla and Fogel (1999), and others? According to Dembski (2002b), "The No Free Lunch theorems show that for evolutionary algorithms to output CSI they had first to receive a prior input of CSI" (223). (By CSI, Dembski means complex specified information, which he uses interchangeably with specified complexity.) In fact, the NFL theorems show nothing of the sort. To see why, let's take a brief excursion into optimization theory.

As an example, consider an expedition to a remote mountainous region. If mountain climbers are interested in finding the highest peak, they have to perform an optimization search over all the peaks—that is, over the physical relief of a mountainous region. They move over the landscape, climb up a mountain at each location, and note its height as measured by an altimeter until they locate the highest peak. Many details of the search for the highest peak are irrelevant to the NFL theorems, including questions such as how to determine that the highest peak has indeed been found and how to know which peak is next in height above the summit already reached. NFL theorems are very general and apply to a wide variety of searches and landscapes; they do not take specific details into account.

Assume that, before embarking on an expedition, we first want to prepare instructions for climbing mountains in an unexplored region. Let us call such sets of instructions mountain-climbing algorithms (MCA). One group of mountaineers suggests starting at a peak located where the path that leads us to the region approaches its periphery. Then the mountaineers will climb peak after peak, moving gradually toward the center of the region, regardless of whether each next peak is higher or lower than the preceding one. We will call the MCA prepared according to such a strategy a center-directed algorithm (CDA).

Another group suggests a different strategy. A CDA, these mountaineers say, might miss the highest peak if it is located away from the center of the region. Their preferred criterion for choosing each next peak for exploration is its being higher than the previously conquered peak. Let us call that MCA a height-oriented algorithm (HOA).

Which of the two algorithms will end the search for the highest peak more quickly? Which algorithm will perform better on the landscape of a particular mountainous region? There is no general answer; it depends on the character of the landscape. On most landscapes, the HOA will perform better than the CDA. If, however, the highest peak is very close to the center of the region and the heights of the surrounding peaks decrease haphazardly toward the periphery, then a search that uses a CDA may outperform an HOA.

Which of the two MCAs is better overall? There are many mountainous countries on earth. Some may have a physical relief wherein the highest point is closest to the geographical center of the region, with a haphazard distribution of lower peaks around the central one. The highest peak of others may be located away from the geographical center, and the heights of the lower peaks may decrease gradually with the distance from that highest peak. There are many other possibilities. Overall, the number of landscapes in which the HOA outperforms the CDA can be expected not to exceed substantially the number of possible landscapes in which CDA is better. In other words, the HOA and the CDA averaged over all possible landscapes can be reasonably expected to perform similarly.

This is a simplified illustration of the NFL theorems. In those theorems Wolpert and Macready (1997) put the simple observation about the equal average performance of various algorithms into a rigorous mathematical form and reveal some subtle features of the algorithms' behavior that are not intuitively evident.

The NFL Theorems: Still with No Mathematics

The mountainous landscape we discussed is a particular case of what is generally called a fitness landscape, and the heights of the peaks in our example is a particular case of a fitness function. Imagine two algorithms conducting a search on a given fitness landscape. They move from point to point over the search space (choosing the search points either at random or in a certain order). After having performed, say, m measurements, an algorithm produces what Wolpert and Macready call a sample—a table wherein the m measured values of the fitness function are listed in temporal order. Generally speaking, two arbitrarily chosen algorithms will not yield identical samples. The probability of algorithm a_1 producing a specific table that is m rows long is different from the probability of algorithm a_2 producing the same table after the same number of iterations.

Enter the first NFL theorem: if the results of the two algorithms' searches are compared not for a specific fitness landscape but averaged over all possible landscapes, the probabilities of obtaining the same sample are equal for any pair of algorithms. The quantity that is averaged is the probability of generating a given sample by an algorithm. This is an exact translation of the first NFL theorem from its mathematically symbolic form into plain words.

The NFL theorems do not restrict the value of m, the number of iterations. There is no condition that the search stops when a certain preselected number of iterations has been completed or when a preselected value of the fitness function has been found. In other words, the concept of a target is absent from the theorems. On the other hand, they do not forbid the algorithms to be target-oriented. The theorems are indifferent to algorithms' having or not having a target.

The NFL theorems are often discussed in terms of algorithms' performance, although the concept of a performance measure is not part of the theorems as such. According to Wolpert and Macready, "the precise way that the sample is mapped to a performance measure is unimportant" (73). The NFL theorems allow for wide latitude in the choice of performance measures. In particular, whereas the theorems themselves do not refer to any target of a search, the algorithms in question may be either target-oriented or not.

Following are examples of targeted and nontargeted algorithms, both equally subject to the NFL theorems. Assume that the fitness function is simply the height of peaks in a specific mountainous region. If we choose a target-oriented algorithm, the target of the search can be defined as a specific peak P whose height is, say, 6000 meters above sea level. In this case the number n of iterations required to reach the predefined height of 6000 meters may be chosen as the performance measure. Then algorithm a_1 performs better than algorithm a_2 if a_1 converges on the target in fewer steps than a_2 does. If two algorithms generated the same sample after m iterations, then they would have found the target—peak P—after the same number n of iterations. The first NFL theorem tells us that the average probabilities of reaching peak P in m steps are the same for any two algorithms. Any two algorithms will have an equal average performance, provided that the averaging is over all possible fitness landscapes (not all of which must in fact exist materially). In the example discussed, the average number n of iterations required to locate the target is the same for any two algorithms, if the averaging is done over all possible mountainous landscapes.

Importantly, the NFL theorems do not say anything about the relative performance of algorithms a_1 and a_2 on a specific landscape. On a specific landscape, either a_1 or a_2 may happen to be much better than its competitor.

Algorithms can also be compared in a targetless context. For example, rather than defining a target as a certain peak P or even as a peak of a certain height, the algorithms may be compared by finding out which of them, a_1 or a_2, finds a higher peak after a certain number m of iterations. The performance measure in this case is the height of a peak reached after m iterations. No specific peak and no specific height is preselected as a target. An algorithm a_1 that after m iterations finds a higher peak than algorithm a_2 performs better. The first NFL theorem tells us that, if averaged over all possible mountainous reliefs (not all of them necessarily existing), the probabilities of both a_1 and a_2 generating the same sample after m iterations are equal. This also means that, in all likelihood, the height of a peak reached after m iterations, if averaged over all possible landscapes, will be the same for any two algorithms.

The NFL theorems are certainly valid for evolutionary algorithms. As Wolpert (2002) reports, Wolpert and Macready have proven recently that the NFL theorems are invalid for coevolutionary algorithms, but that is a different question.

The NFL Theorems: A Little Mathematics

A series of NFL theorems pertains to various situations. The original theorems (Wolpert 1996a, 1996b) dealt with problems of supervised learning. Later they were extended to optimization problems associated with search algorithms (Wolpert and Macready 1997).

The first NFL theorem for search pertains to fixed (time-independent) fitness landscapes, while the second is for time-dependent landscapes. Although there is a substantial difference between the two, their principal meaning can be understood by reviewing only the first.

Imagine a finite set X called the search space and a fitness function f that assigns a value to each point of X; the values of f are within a range denoted Y. Altogether the search points and their fitness values form the fitness landscape. We consider algorithms that explore X one point at a time. At each step, the algorithm decides which point to examine next, depending on the points that have been examined already and their fitness values, but it does not know the fitness of any other points. This decision might even be made at random or partly at random.

After an algorithm has iterated the search m times, we have a time-ordered set (a sample) denoted d_m^Y, which comprises m measured values of f within the range Y. Let P be the conditional probability of having obtained a given sample d_m^Y after m iterations for given f, Y, and m. Then, according to the first NFL theorem,

$$\sum_f P(d_m^Y | f, m, Y, a_1) = \sum_f P(d_m^Y | f, m, Y, a_2) \qquad (1)$$

where a_1 and a_2 are two different algorithms. The summation is performed over all possible fitness functions. Equation (1) means that, in probabilistic terms, the results of a search, if averaged over all possible fitness landscapes, are the same for any pair of algorithms.

The equation for the second NFL theorem—for time-dependent landscapes—differs in two respects. First, it contains one more factor affecting the algorithm's behavior: an evolution operator, which is a rule reflecting how the landscape evolves from iteration to iteration. Second, the probabilities of obtaining a given sample are averaged over all possible evolution operators rather than over all possible fitness functions. For the purpose of this chapter, it is sufficient to refer to the first NFL theorem only.

Equation (1) also says that, if the performance of an algorithm a_1 is superior to that of another algorithm a_2 when applied to a specific class of fitness functions, it is necessarily inferior to the performance of a_2 on some other class of fitness functions. (In my previous analogy the mountain-climbing algorithm denoted COA performed better than the HOA on one type of physical relief but worse than the HOA on another.)

Algorithms do not incorporate any prior knowledge of the properties of the fitness function. They are therefore called black-box algorithms. They operate on a fitness landscape without prior knowledge of the landscape's relief, probing point after point, either deterministically or stochastically (according to a rule which has some random component).

Importantly, the NFL theorems do not say anything about the performance of any two algorithms on any particular landscape. If, in equation (1), we remove the summation symbols, the sign of equality must be replaced with an inequality. In other words, except for rare special cases,

$$P(d_m^Y \mid f, m, Y, a_1) \neq P(d_m^Y \mid f, m, Y, a_2) \qquad (2)$$

Inequality (2) means that, generally, the performance of any two arbitrarily chosen algorithms on a specific landscape cannot be expected to be equal.

The NFL theorems do not address a situation wherein a certain algorithm a_1 significantly outperforms algorithm a_2 on a few landscapes while there are no such landscapes where a_2 is much better than a_1. According to the theorems, in such cases a_2 outperforms a_1 on many landscapes, but only slightly. This situation (known technically as head-to-head minimax asymmetry) can be defined rigorously in quantitative terms (Wolpert and Macready 1997, 74).

As Wolpert and Macready point out, "there is always a possibility of asymmetry between algorithms if one of them is stochastic" (76). The asymmetry may be more significant than the equal average performance of algorithms established by the NFL theorems. This point is relevant to evolutionary algorithms, including Darwinian genetic algorithms and Dembski's analysis of them. Dembski (2002b) states that "an evolutionary algorithm is a stochastic process" (189). For stochastic algorithms, the possibility of the minimax asymmetries is real, and when such asymmetries arise, they make the NFL theorems practically irrelevant.

Here is how Dembski defines the NFL theorems: "A generic NFL theorem now takes the following form: It sets up a performance measure M that characterizes how effectively an evolutionary algorithm E locates a target T within m steps using information j." According to him, information j resides in an "information-resource space," which is beyond the "phase space" (his term for the search space) and usually exceeds the search space in size and complexity (200–3).

As follows from our preceding discussion, Dembski's definition misrepresents the NFL theorems. They do not set performance measures but only compare the generation of samples by algorithms within m iterations. Performance measures are introduced within the framework of corollaries to and interpretations of the theorems and can be chosen in variety of ways. They should match the outcome of the search. Furthermore, there is no concept of a target (for which Dembski offers no definition) in the NFL theorems as such; they are valid for targetless searches as well. (Each search is supposed to lead to some outcome but not necessarily to a target. An outcome is a general, qualitative concept; it may be either intended or not. It is not necessarily connected to the termination of a search; a search may be terminated for reasons unrelated to the outcome. A target is a specific, often quantitative concept, such as a predefined value of the fitness function that terminates the search when it is found.)

Moreover, there is no talk about information *j* in the parlance of the NFL theorems. These theorems are about black-box algorithms, which start a search without prior information about the fitness landscapes but continue (and complete, if appropriate) the search using the information they extract gradually from the fitness function in the course of the search. This information is sufficient to continue a search at every step. Contrary to Dembski, the search algorithms do not need to go for information into a higher-order information-resource space.

Continuing, Dembski writes, "since blind search always constitutes a perfectly valid evolutionary algorithm, this means that the average performance of any evolutionary algorithm E is no better than blind search" (202). This is correct, but the word *average* is crucial. Dembski forgets this word when he interprets the NFL theorems as making it impossible for evolutionary algorithms to outperform blind search on specific landscapes. On the contrary, the theorems do not assert that no evolutionary algorithm performs better than a random sampling or a blind search. Such a statement is valid only for the performance of algorithms if evaluated on average for all classes of problems. It is invalid when specific genetic pathways are considered.

As Wolpert and Macready (1997) emphasize, the NFL theorems do not predict performance in the real world. In fact, the uniform average is a crude tool designed to analyze the relationship between search algorithms and fitness functions. Wolpert (2002) has pointed out that, in Dembski's discourse, the factors arising in the NFL theorems "are never specified in his analysis." According to Wolpert,

> throughout Dembski's discourse there is a marked elision of the formal details of the biological processes under consideration. Perhaps the most glaring example of this is that neo-Darwinian evolution of ecosystems does not involve a set of genomes all searching the same,

fixed fitness function, the situation considered by the NFL theorems. Rather it is a coevolutionary process. Roughly speaking, as each genome changes from one generation to the next, it modifies the surfaces that the other genomes are searching. (n.p.)

Furthermore, recent results (Wolpert 2002) indicate that "NFL results do not hold in coevolution" (n.p.).

The Displacement Problem

In his response to one of his critics (Orr 2002), Dembski (2002a) says, "Given my title, it's not surprising that critics see my book *No Free Lunch* as depending crucially on the No Free Lunch theorems of Wolpert and Macready. But in fact, my key point concerns displacement, and the NFL theorems merely exemplify one instance (not the general case)" (n.p.).

In fact, however, Dembski (2002b) introduces the displacement problem in the section on the NFL theorem (200–3) as a consequence of his interpretation of these theorems: "The significance of the NFL theorems is that an information-resource space J does not, and indeed cannot, privilege a target T" (202). He introduces two concepts here: a target and an information-resource space J. In fact, the significance of the NFL theorems can hardly be seen in the quoted statement. As we have discussed, the concept of a target as such is absent from the NFL theorems. They are equally valid for targeted and targetless searches. Nor is there any talk in the theorems about information-resource spaces.

Here's how Dembski introduces the displacement problem: "the problem of finding a given target has been displaced to the new problem of finding the information j capable of locating that target. Our original problem was finding a certain target within phase space. Our new problem is finding a certain j within the information-resource space J" (203). This quotation contains arbitrary assertions. First, the NFL theorems contain nothing about any arising information-resource space. If Dembski wanted to introduce that concept within the framework of the theorems, at the very least, he should have shown what the role of an information-resource space is in view of the black-box nature of the algorithms in question. Second, the NFL theorems are indifferent to the presence or absence of a target in a search, which alone leaves Dembski's introduction of the displacement problem, with its constant references to targets, hanging in the air.

The Irrelevance of the NFL Theorems

I submit that the real question is not whether or not the NFL theorems are valid for evolutionary algorithms (EA). Within the scope of their legitimate interpretation—when the conditions assumed for their derivation hold—the theorems certainly apply to EAs. The problem arises when they are applied where the assumed premises do not hold. Although Wolpert and Macready have shown that the theorems may not hold in the case of coevolution, my conclusion would not change even if they were also valid for coevolution.

The simple fact is that the NFL theorems are irrelevant to the real question we face. This is the actual, two-tiered question:

1. Can an evolutionary algorithm outperform random sampling (or blind search) in situations of interest?
2. Can specified complexity be purchased without intelligence?

Dembski's answer to part 1 is a categorical no. He is wrong. The correct answer is a categorical yes. Let me show why. Dembski's no is partially based on the alleged mathematical certainty expressed by the NFL theorems, according to which no algorithm performs better than a random search does. Indeed, we read, "The No Free Lunch theorems dash any hope of generating specified complexity via evolutionary algorithms" (Dembski 2002b, 196). He also tells us that "The No Free Lunch theorems show that evolutionary algorithms, apart from careful fine-tuning by a programmer, are no better than blind search and thus no better than pure chance" (212).

What Dembski seems to ignore is the crucial point that I have stressed several times: the NFL theorems legitimately compare the performance of any two algorithms, but what they compare is performance averaged over all possible fitness landscapes. This is an interesting theoretical conclusion and a tool for investigating the mutual relationship between the fitness functions and search algorithms. It has no relevance for problems of practical interest encountered in real life, where we are interested in finding out whether or not a given algorithm outperforms a random search if applied to a specific class of fitness landscapes.

There are plenty of examples showing that evolutionary algorithms indeed outperform random search when applied to fitness functions of interest. Let us recall some of them. In Dawkins's example, as Dembski tells us himself, a random search is expected to converge on the target phrase after about 10^{40} iterations. If, however, Dawkins's evolutionary algorithm is applied to the same task, it achieves the same result after about only forty iterations. Even if Dawkins's algorithm is replaced by Dembski's version, it will reach the target, as Dembski says, after about 4000 iterations. 4000 versus 10^{40}: this is out-performance and a very respectable outperformance indeed. What significance has the fact that the algorithms cannot outperform a random search if aver-aged over all possible fitness functions? They outperform a random search if applied to the specific fitness functions of interest, and that is all that counts.

In the cases of a search for the optimal shape of an antenna (Altshuler and Linden 1999) or of a checkers-playing algorithm (Chellapilla and Fogel 1999) both of which Dembski (2002b, 221) views more favorably than he does Dawkins's algorithm, again the evolutionary algorithms immensely outperform a random search. Although the NFL theorems are valid for Altshuler and Linden's and Chellapilla and Fogel's algorithms, this fact is of no consequence because what those authors are interested in is not the averaged performance over all possible fitness functions but the performance on a specific class of fitness functions; and the NFL theorems say nothing about such performance.

Rather than equation (1), in practical situations, inequality (2) is really relevant; it says that different algorithms perform differently on specific classes of fitness functions. Hence, Dembski's discussion of the NFL theorems is of no consequence for the question of whether evolutionary algorithms can out-perform a random search. They can and they do.

Of course, Dembski has an escape clause: He admits that evolutionary algorithms can outperform a random sampling if there is "careful fine-tuning by a programmer" (212). If, however, a programmer can design an evolutionary algorithm that is fine-tuned to ascend certain fitness landscapes, what can pro-hibit a naturally arising evolutionary algorithm to fit in with the kinds of land-scape it faces? Nothing can, and nothing does. If a specific evolutionary algorithm, either fine-tuned by a programmer or arising naturally, outperforms random sampling on a specific landscape, the NFL theorems are of no consequence, and Dembski's reference to these theorems is irrelevant.

This thesis can be illustrated as follows: naturally arising fitness landscapes will frequently have a central peak topping relatively smooth slopes. If a cer-tain property of an organism, such as its size, affects the organism's survivabil-ity, then there must be a single value of the size most favorable to the organism's fitness. If the organism is either too small or too large, its survival is at risk (Haldane 1928, 20–28). If there is an optimal size that ensures the highest fitness, then the relevant fitness landscape must contain a single peak of the highest fitness surrounded by relatively smooth slopes.

The graphs in figure 11.1 schematically illustrate my thesis. The fitness function may be, for example, the average life expectancy of an animal, the average number of its surviving descendants, or some other single-valued quan-tity that reflects an animal's success at survival. The fitness function is repre-

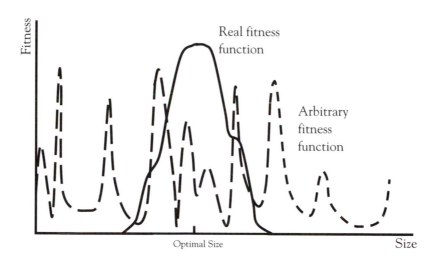

Figure 11.1. Fitness as a function of some characteristic, in this case the size of an animal. Solid curve: schematic representation of a naturally arising, single-valued fitness function, wherein the maximum fitness is achieved for a certain optimal size. Dashed curve: an imaginary, rugged fitness function.

sented by the solid curve, which has a well-defined peak corresponding to the optimal size, with more or less smooth slopes on both sides of the peak. We may imagine many other possible fitness functions, such as the rugged fitness function represented by the dashed curve. Such fitness functions, however, do not represent biological reality: the survivability of an animal cannot depend on its size (or on some other feature) in such a haphazard manner. It is unlikely that several different sizes will be comparably advantageous to an organism's fitness.

Obviously, in this case, the evolutionary algorithm based on natural selection is well suited to ascending the actual fitness landscape. Indeed, the closer the organism's size to the maximum of the fitness landscape, the more it is favored by natural selection. In this case the NFL theorems, while they are correct (if we ignore coevolution), are irrelevant. Natural selection will perform well on the actual landscape, certainly better than a random search. Other algorithms can perform better on other possible fitness landscapes, such as the rugged landscape exemplified in figure 11.1, but landscapes actually encountered in the biosphere are not likely to be so rugged. (This example assumes a one-dimensional fitness function, whereas a real fitness function is multidimensional; the example is intended only to illustrate the point.)

The debate can be extended, as Stuart Kauffman (2000) has done. If Darwinian evolution has been indeed taking place, obviously Darwinian evolutionary algorithms work well on the fitness landscapes that arise naturally in the biosphere. Then, according to the NFL theorems, these algorithms must perform poorly on some other possible fitness landscapes. In other words, while the natural evolutionary algorithms entailing random mutations and natural selection (plus recombination and possibly other mechanisms) do indeed outperform a random search, they should underperform a random search on different fitness landscapes that could have existed in some alternative reality. So why, of all the enormous variety of possible fitness landscapes, are the fitness landscapes actually observed in the biosphere exactly those accessible to Darwinian evolutionary algorithms? The answer is that, because of the enormous variety of possible evolutionary algorithms and fitness functions, the probability of some fraction of algorithms being naturally fine-tuned to the existing landscapes is close to certainty. The example with the fitness function that depended on an organism's size illustrates this statement.

Posing Kauffman's question shifts the discussion from the relevance of the NFL theorems for the observed biological reality to the realm of anthropic coincidences (see chapter 12 in this book). Whatever the explanation of those coincidences may be (Drange 1998; Stenger 2001a, 2001b; Ikeda and Jefferys 2001; Perakh 2001a), it does not alter the conclusion that the NFL theorems are irrelevant to the comparison of evolutionary algorithms with a random search as long as we discuss existing biological reality.

Dembski's answer to part 2 of my question, "Can specified complexity be purchased without intelligence?" is also a categorical no. To my mind, yes is more plausible. The necessity of intelligence for generating SC is something Dembski ostensibly sets out to prove; in fact, he often *uses it as a given*. He provides no evidence that would meet the requirements of scientific rigor to substantiate his thesis but only arbitrary assumptions lacking evidence. When he attempts to apply more-specific arguments, such as those based on the NFL theorems and its alleged implication in the form of the displacement problem, his discourse is contradictory and inconsistent. On the other hand, evolutionary biologists have suggested plausible scenarios explaining how evolutionary algorithms can work without the interference of an external intelligence.

ID advocates often charge that such scenarios are just-so stories and therefore unconvincing. There is, however, often substantial empirical evidence in favor of these scenarios, and the biological literature abounds in them. Moreover, such a reproach sounds odd coming from ID advocates, whose entire conceptual system is a just-so story in which a blanket reference to "intelligent design" is nothing more than a Dembskian Z-factor (see chapter 13 in this book) offered as a substitute for a realistic scenario.

The Displacement "Problem"

Dembski (2002a) asserts that the displacement problem is, in fact, the core of his thesis. At a close inspection, however, it becomes clear that the displacement problem is irrelevant to real-life situations. Recall that he defines it as "the problem of finding a given target . . . displaced to the new problem of finding the information j capable of locating that target. Our original problem was finding a certain target within phase space. Our new problem is finding a certain j within the information-resource space J" (Dembski 2002b, 203). As he explains, "the fitness function is of course the additional information that turns the blind search to a constrained search" (202). Hence, the information-resource space J is meant by Dembski as a space of (possibly along with other sources of information) all possible fitness functions.

According to Dembski, the information-resource space J is "in practice . . . much bigger and much less tractable than the original phase space" (203). Hence, the original problem has been displaced to a much more intractable problem. To solve the new problem, he insists, the specified complexity must be injected by intelligence. In summary, his displacement problem means that the space of all possible fitness functions has to be searched to determine the fitness function for the problem at hand.

Dembski gives us no reason to assume that the information-resource space is much larger and much less tractable than the original phase space. In fact, there seem to be no such reasons. The information-resource space can be larger, about the same size, or smaller than the phase space. Dembski provides an example of a search for a treasure buried on an island (204). Instead of a search all over the island (whose topography constitutes the phase space), the search may be displaced to a worldwide search for a map of the island, wherein the location of the treasure is indicated. Now the information-resource space is the entire globe, which is immensely larger than the island in question.

This example can easily be reversed since it could happen as well that finding the map in question is much easier than finding the treasure itself without a map. Indeed, if it is known that the map is hidden in a certain building in a certain city, the information-resource space becomes the specific building and is much smaller and much more tractable than the original phase space (which was the entire island). But regardless of which space is larger and less tractable, and regardless of the very existence or absence of the displacement problem, it is irrelevant for a real-life optimization search. Here is why.

To start a search, a black-box algorithm needs no information about the fitness function. To continue the search, an algorithm needs information from the fitness function, but no search of the space of all possible fitness functions is needed. In the course of a search, the algorithm extracts the necessary information from the landscape it is exploring. The fitness landscape is always given and automatically supplies sufficient information to continue and complete the search.

Consider Dawkins's WEASEL algorithm. It explores the available phrases and selects from them, using the comparison of the intermediate phrases with the target. The fitness function has, in this case, built-in information necessary to perform the comparison. This fitness function is given to the search algorithm; to provide this information to the algorithm, no search of a space of all possible fitness functions is needed and therefore is not performed.

The same is true for natural evolutionary algorithms. The evolutionary algorithms, both designed by intelligence and occurring spontaneously, deal with given, specific fitness functions and have no need to search the information-resource space. Dembski's displacement problem is a phantom.

Conclusion

Dembski (1998d) has written, "As Christians we know that naturalism is false" (14). Obviously, if one "knows" something, this ends a discussion. Since 1998, his attitude does not seem to have changed. Recently (2002d), he asserted that the ID advocates will never capitulate to their detractors. If so, then his statement testifies to a fact noted by critics of the ID "theory": ID is not science. Scientists normally admit that, no matter what theories are commonly accepted at any time, there is always a chance they may be overturned by new evidence. Genuine scientists would not make statements about never capitulating to their detractors, no matter what.

The following points encapsulate the gist of this chapter:

- Dembski's critique of Dawkins's targeted evolutionary algorithm fails to repudiate the illustrative value of Dawkins's example, which demonstrates how supplementing random changes with a suitable law increases the rate of evolution by many orders of magnitude.
- Dembski ignores Dawkins's targetless evolutionary algorithm, which successfully illustrates spontaneous increase of complexity in an evolutionary process.
- Contrary to Dembski's assertions, evolutionary algorithms routinely outperform a random search.
- Contrary to Dembski assertion, the NFL theorems do not make Darwinian evolution impossible. Dembski's attempt to invoke the theorems to

prove otherwise ignores the fact that they assert the equal performance of all algorithms only if averaged over all fitness functions.

- Dembski's constant references to targets when he discusses optimization searches are based on his misinterpretation of the NFL theorems, which entail no concept of a target. Moreover, his discourse is irrelevant to Darwinian evolution, which is targetless.
- The so-called displacement problem, touted by Dembski as the core of his thesis, is a phantom because evolutionary algorithms face given, specific fitness landscapes. The landscape supplies sufficient information to continue and (when appropriate) complete a search; there is no need to search the higher-order information-resource space.
- The question "Why are the evolutionary algorithms actually observed in the biosphere well adjusted to the actually observed fitness functions?" belongs in the general discussion of anthropic coincidences. The arguments showing that the anthropic coincidences do not require the hypothesis of a supernatural intelligence also answer the questions about the compatibility of fitness functions and evolutionary algorithms.

Acknowledgments

I had the privilege of advice from David H. Wolpert, co-author of the no-free-lunch theorems, who shared ideas pertaining both to the essence of the theorems and to Dembski's treatment of them. I also appreciate comments by Taner Edis, Gordon Elliott, Thomas D. Schneider, Jeffrey Shallit, Erik Tellgren, Matt Young, and especially Brendan McKay. Of course, the opinions and arguments in this chapter, and even more so possible errors, are mine.

Chapter 12

Is the Universe Fine-Tuned for Us?

VICTOR J. STENGER

THE ANCIENT ARGUMENT from design for the existence of God is based on the common intuition that the universe and life are too complex to have arisen by natural means alone. As philosopher David Hume pointed out in the eighteenth century, however, the fact that we cannot explain some phenomenon naturally does not allow us to conclude that it had to be a miracle.

In recent years, novel versions of the argument from design that call upon modern science as their authority have appeared on the scene. Proponents of so-called intelligent design confidently claim to rule out natural processes as the sole origin for certain biological systems (Behe 1996; Dembski 1998a, 1999, 2002b). Here I focus on another variation of the argument from design—the argument from fine tuning, in which evidence for a purposeful creation is seen in the laws and constants of physics.

This claim of evidence for a divine cosmic plan is based on the observation that earthly life is so sensitive to the values of the fundamental physical constants and properties of its environment that even the tiniest changes to any of these would mean that life, as we see it around us, would not exist. The universe is then said to be exquisitely fine-tuned: delicately balanced for the production of life. As the argument goes, the chance that any initially random set of constants would correspond to the set of values that we find in our universe is very small, and the universe is exceedingly unlikely to be the result of mindless chance. Rather, an intelligent, purposeful, and indeed caring personal creator must have made things the way they are.

Some who make the fine-tuning argument are content to suggest merely that intelligent, purposeful, supernatural design has become an equally viable alternative to the random, purposeless, natural evolution of the universe and humankind suggested by conventional science. This mirrors recent arguments for intelligent design as an alternative to evolution.

A few design advocates, however, have gone further to claim that God is now *required* by scientific data. Moreover, this God must be the God of the Christian Bible. They insist that the universe is provably not the product of purely natural, impersonal processes. Typifying this view is physicist and astronomer Hugh Ross (1995), who cannot imagine fine tuning happening any other way than by a "personal Entity . . . at least a hundred trillion times more 'capable' than are we human beings with all our resources." He concludes that "the Entity who brought the universe into existence must be a Personal Being, for only a person can design with anywhere near this degree of precision" (118).

The delicate connections among certain physical constants and between those constants and life I will collectively call *anthropic coincidences*. Before examining the merits of the interpretation of these coincidences as evidence for intelligent design, I will review how the notion first came about. John Barrow and Frank Tipler (1986) provide a detailed history, a wide-ranging discussion of all the issues, and a complete list of references. But be forewarned that their exhaustive tome has many errors, especially in equations, some of which remain uncorrected in later editions.

The Large-Number Coincidences

Early in the twentieth century, Hermann Weyl (1919) expressed his puzzlement that the ratio of the electromagnetic force to the gravitational force between two electrons is such a huge number: $N_1 = 10^{39}$. This means that the strength of the electromagnetic force is greater than the strength of the gravitational force by 39 orders of magnitude. Weyl puzzled over this, expressing his intuition that pure numbers like π, which occur in the description of physical properties, should most naturally occur within a few orders of magnitude of 1. You might expect the numbers 1 or 0 naturally. But why 10^{39}? Why not 10^{57} or 10^{-123}? Some principle must select out 10^{39}, according to Weyl's way of thinking.

Sir Arthur Eddington (1923) observed further, "It is difficult to account for the occurrence of a pure number (of order greatly different from unity) in the scheme of things; but this difficulty would be removed if we could connect it to the number of particles in the world—a number presumably decided by accident" (167). He estimated that number, now called the Eddington number, to be $N = 10^{79}$. Well, N is not too far from the square of N_1.

Look around at enough numbers and you are bound to find some that appear to be connected. Most physicists, then and now, do not regard the large-numbers puzzle seriously. It seems like numerology. But the great physicist Paul Dirac (1937) noticed that N_1 is the same order of magnitude as another pure number N_2, which gives the ratio of a typical stellar lifetime to the time for light to traverse the radius of a proton. That is, he found two seemingly unconnected large numbers to be of the same order of magnitude. If one number being large is unlikely, how much more unlikely is another to come along with about the same value?

Robert Dicke (1961) pointed out that N_2 is necessarily large in order that the lifetime of typical stars be sufficient to generate heavy chemical elements such as carbon. Furthermore, he showed that N_1 must be of the same order as N_2 in any universe with heavy elements. Bernard Carr and Martin Rees (1979) picked up the argument, claiming to show that the orders of magnitudes of masses and lengths at every level of structure in the universe are fixed by the values of just three constants: the dimensionless strengths of the electromagnetic and gravitational forces and the electron-proton mass ratio.

Making Carbon

The heavy elements did not get fabricated straightforwardly. According to the big-bang theory, only hydrogen, deuterium (the isotope of hydrogen consisting of one proton and one neutron), helium, and lithium were formed in the early universe. Carbon, nitrogen, oxygen, iron, and the other elements of the chemical periodic table were not produced until billions of years later. These billions of years were needed for stars to form and, near the end of their lives, assemble the heavier elements out of neutrons and protons. When the more massive stars expended their hydrogen fuel, they exploded as supernovae, spraying the manufactured elements into space. Once in space, these elements cooled, and gravity formed them into planets.

Billions of additional years were needed for our home star, the sun, to provide a stable output of energy so at least one of its planets could develop life. But if the gravitational attraction between protons in stars had not been many orders of magnitude weaker than the electric repulsion, as represented by the very large value of N_1, stars would have collapsed and burned out long before nuclear processes could build up the periodic table from the original hydrogen and deuterium. The formation of chemical complexity is likely only in a universe of great age.

Great age is not all. The element-synthesizing processes in stars depend sensitively on the properties and abundances of deuterium and helium produced in the early universe. Deuterium would not exist if the difference between the masses of a neutron and a proton were just slightly displaced from its actual value. The relative abundances of hydrogen and helium also depend strongly on this parameter. They, too, require a delicate balance of the relative strengths of gravity and the weak force—the force responsible for nuclear beta decay. A slightly stronger weak force, and the universe would be 100 percent hydrogen; all the neutrons in the early universe would have decayed, leaving none around to be saved in deuterium nuclei for later use in the synthesizing elements in stars. A slightly weaker weak force, and few neutrons would have decayed, leaving about the same numbers of protons and neutrons; then all the protons and neutrons would have been bound up in helium nuclei, with two protons and two neutrons in each. This would have led to a universe that was 100 percent helium, with no hydrogen to fuel the fusion processes in stars. Neither of these extremes would have allowed for the existence of stars and life as we know it based on carbon chemistry (Livio et al. 1989).

The electron also enters into the tightrope act needed to produce the heavier elements. Because the mass of the electron is less than the neutron-proton mass difference, a free neutron can decay into a proton, an electron, and an anti-neutrino. If the mass of the electron were just a bit larger, the neutron would be stable, and most of the protons and electrons in the early universe would have combined to form neutrons, leaving little hydrogen to act as the main component and fuel of stars. The neutron must also be heavier than the proton, but not so much heavier that neutrons cannot be bound in nuclei.

In 1952, astronomer Fred Hoyle (1954) used anthropic arguments to predict that an excited carbon nucleus has an excited energy level at around 7.7 megaelectronvolts (MeV). The success of this prediction gave credibility to anthropic reasoning, so let me discuss this example in detail, because it is the only successful prediction of this line of inference so far.

I have already noted that a delicate balance of physical constants was necessary for carbon and other chemical elements beyond lithium in the periodic table to be cooked in stars. Hoyle looked closely at the nuclear mechanisms involved and found that they appeared to be inadequate.

The basic mechanism for the manufacture of carbon is the fusion of three helium nuclei into a single carbon nucleus:

$$3\mathrm{He}^4 \rightarrow \mathrm{C}^{12} \tag{1}$$

(The superscripts give the number of nucleons—that is, protons and neutrons in each nucleus—which is specified by its chemical symbol. The total number of nucleons is conserved—that is, remains constant—in a nuclear reaction.) The probability of three bodies coming together simultaneously is very low, however, and some catalytic process in which only two bodies interact at a time must be assisting. An intermediate process in which two helium nuclei first fuse into a beryllium nucleus, which then interacts with the third helium nucleus to give the desired carbon nucleus, gives the desired result:

$$2He^4 \rightarrow Be^8 \tag{2}$$

$$He^4 + Be^8 \rightarrow C^{12} \tag{3}$$

Hoyle (1954) showed that this still was not sufficient unless the carbon nucleus had a resonant excited state at 7.7 MeV to provide for a high reaction probability. A laboratory experiment was undertaken, and sure enough: a previously unknown excited state of carbon was found at 7.66 MeV (Hoyle et al. 1953).

Nothing can gain you more respect in science than the successful prediction of an unexpected new phenomenon. Here, Hoyle used standard nuclear theory. But his reasoning contained another element whose significance is still hotly debated. Without the 7.7-MeV nuclear state of carbon, our form of life based on carbon would not have existed.

The Anthropic Principles

Like the large-number coincidences, the 7.7-MeV nuclear state seems unlikely to be the result of chance. The existence of these apparent numerical coincidences led Brandon Carter (1990) to introduce the notion of an anthropic principle, which hypothesizes that the coincidences are not accidental but somehow built into the structure of the universe. Barrow and Tipler (1986) have identified three different forms of the anthropic principle, defined as follows: "Weak anthropic principle (WAP): The observed values of all physical and cosmological quantities are not equally probable but take on values restricted by the requirement that there exist sites where carbon-based life can evolve and by the requirement that the universe be old enough for it to have already done so." (21). The WAP merely states the obvious. If the universe was not the way it is, we would not be the way we are. But it is sufficient for predictions such as Hoyle's.

Barrow and Tipler continue: "Strong anthropic principle (SAP): The universe must have those properties which allow life to develop within it at some stage in its history" (21). This is essentially the form originally proposed by Carter, which suggests that the coincidences are not accidental but the result of a law of nature. It is a strange law indeed, unlike any other in physics. It suggests that life exists as some Aristotelian final cause, as has been suggested by the proponents of intelligent design.

Barrow and Tipler (1986, 22) argue that the SAP can have three interpretations:

1. There exists one possible universe "designed" with the goal of generating and sustaining "observers" (the interpretation adopted by most design advocates).
2. Observers are necessary to bring the universe into being (a form of solipsism found in today's new-age quantum mysticism).
3. An ensemble of other different universes is necessary for the existence of our universe.

This last speculation is part of contemporary cosmological thinking, as I will discuss later in the chapter. It represents the idea that the coincidences are accidental. We just happen to live in the particular universe that was suited for us. The current dialogue focuses on the choice between interpretations 1 and 3; item 2 is not taken seriously in the scientific and theological communities (Stenger 1995).

Before discussing the relative merits of the three choices, let me complete the story on the various forms of the anthropic principle discussed by Barrow and Tipler. In addition to the two anthropic principles already mentioned, they identify another version: "Final anthropic principle (FAP): Intelligent information-processing must come into evidence in the universe, and, once it comes into existence, it will never die out" (21). Martin Gardner (1986) referred to this as the "completely ridiculous anthropic principle (CRAP)" (22–25).

Interpreting the Coincidences

Many religious thinkers see the anthropic coincidences as evidence for a purposeful design of the universe. They ask, How can the universe possibly have obtained the unique set of physical constants it has, so exquisitely fine-tuned for life as they are, except by purposeful design—design with life and perhaps humanity in mind (Swinburne 1998, Ellis 1993, Ross 1995)?

Let us examine the implicit assumptions here. First and foremost, and fatal to the design argument all by itself, is the wholly unwarranted assumption that only one type of life is possible—the particular form of carbon-based life we have here on Earth.

Carbon seems to be the chemical element best suited to act as the building block for the complex molecular systems that develop lifelike qualities. Even today, new materials assembled from carbon atoms exhibit remarkable, unexpected properties, from superconductivity to ferromagnetism. But to assume that only carbon life is possible is tantamount to "carbocentrism," which results from the fact that you and I are structured on carbon.

Given the known laws of physics and chemistry, we can easily imagine life based on silicon (computers, the Internet?) or other elements chemically similar to carbon. These still require cooking in stars and thus a universe old enough for star evolution. The $N_1 = N_2$ coincidence would still hold in this case, although the anthropic principle would have to be renamed the cyberthropic principle or some such, with computers rather than humans, bacteria, and cockroaches the purpose of existence.

Only hydrogen, helium, and lithium were synthesized in the early big bang. They are probably chemically too simple to be assembled into diverse structures. So it seems that any life based on chemistry would require an old universe, with long-lived stars producing the needed materials. Still, we cannot rule out forms of matter other than molecules as building blocks of complex systems. While atomic nuclei, for example, do not exhibit the diversity and complexity seen in the way in which atoms assemble into molecular structures, perhaps they might be able to do so in a universe with different properties and laws.

Sufficient complexity and long life may be the only ingredients needed for a universe to have some form of life. Those who argue that life is highly improbable need to open their minds to the possibility that life might be likely with many different configurations of laws and constants of physics. Furthermore, nothing in anthropic reasoning indicates any special preference for human life or indeed intelligent or sentient life of any sort—just an inordinate fondness for carbon.

Michael Ikeda and William Jefferys (2001) have demonstrated these logical flaws and others in the fine-tuning argument with a formal probability analysis. They have also noted an amusing inconsistency that shows how promoters of design often use mutually contradictory logic: on the one hand, the creationists and God-of-the-gaps evolutionists argue that nature is too uncongenial for life to have developed totally naturally; therefore, supernatural input must have occurred. On the other hand, the fine-tuners (often the same people) argue that the constants and laws of nature are exquisitely *congenial* to life; therefore, they must have been supernaturally created. They can't have it both ways.

How Fine-Tuned Anyway?

Someday we may have the opportunity to study different forms of life that evolved on other planets. Given the vastness of the universe and the common observation of supernovae in other galaxies, we have no reason to assume life exists only on earth. Although it hardly seems likely that the evolution of DNA and other details were exactly replicated elsewhere, carbon and the other elements of our form of life are well distributed throughout the universe, as evidenced by the composition of cosmic rays, meteors, and the spectral analysis of interstellar gas.

We also cannot assume that life would have been impossible in our universe had the physical laws been different. Certainly we cannot speak of such things in the normal scientific mode in which direct observations are described by theory. But at the same time, it is not illegitimate, not unscientific, to examine the logical consequences of existing theories that are well confirmed by data from our own universe.

The extrapolation of theories beyond their normal domains can turn out to be wildly wrong. But it can also turn out to be spectacularly correct. The fundamental physics learned in earthbound laboratories has proved to be valid at great distances from the earth and at times long before the earth and solar system had been formed. Those who argue that science cannot talk about the early universe or life on the early earth because no humans were there to witness these events greatly underestimate the power of scientific theory.

I have made a modest attempt to obtain some feeling for what a universe with different constants would be like. W. H. Press and Alan Lightman (1983) have shown that the physical properties of matter, from the dimensions of atoms to the order of magnitude of the lengths of the day and the year, can be estimated from the values of just four fundamental constants. (This analysis differs slightly from Carr and Rees [1979]). Two of these constants are the strengths of the electromagnetic and strong nuclear interactions. The other two are the masses of the electron and proton. Although the neutron mass does not enter into these calculations, it would still have to have a limited range for there to be neutrons in stars, as already discussed.

I find that long-lived stars, which could make life more likely, will occur over a wide range of these parameters (Stenger 1995, 2000). For example, if we take the electron and proton masses to be equal to their values in our universe, an electromagnetic force any stronger than its value in our universe will give a stellar lifetime of more than 680 million years. The strength of the strong interaction does not enter into this calculation. If we had an electron mass 100,000 times lower, the proton mass could be as much as 1000 times lower to achieve the same minimum stellar lifetime. This is hardly fine tuning.

Many more constants are needed to fill in the details of our universe. And our universe, as we have seen, might have had different physical laws. We have little idea what those laws might be; all we know are the laws we have. Still, varying the constants that go into our familiar equations will give many universes that do not look a bit like ours. The gross properties of our universe are determined by these four constants, and we can vary them to see what a universe might grossly look like with different values of these constants.

I have analyzed 100 universes in which the values of the four parameters were generated randomly from a range five orders of magnitude above to five orders of magnitude below their values in our universe—that is, over a total range of ten orders of magnitude (Stenger 1995, 2000). Over this range of parameter variation, N_1 is at least 10^{33} and N_2 at least 10^{20} in all cases. That is, both are still very large numbers. Although many pairs do not have N_1 = N_2, an approximate coincidence between these two quantities is not very rare.

I have also examined the distribution of stellar lifetimes for these same 100 universes (Stenger 1995, 2000). While a few are low, most are probably high enough to allow time for stellar evolution and heavy-element nucleo-synthesis. Over half the universes have stars that live at least a billion years. Long stellar lifetime is not the only requirement for life, but it certainly is not an unusual property of universes.

I do not dispute that life *as we know it* would not exist if any one of several of the constants of physics were just slightly different. Additionally, I cannot prove that some other form of life is feasible with a different set of constants. But anyone who insists that our form of life is the only one conceivable is making a claim based on no evidence and no theory.

Fine Tuning the Cosmological Constant

Let me discuss an example of supposed fine tuning that arises out of cosmology. This is the apparent fine tuning of Albert Einstein's cosmological constant within 120 orders of magnitude, without which life would be impossible. This will require some preliminary explanation.

When Einstein first wrote down his equations of general relativity in 1915, he saw that they allowed for the possibility of gravitational energy stored in the curvature of empty space-time. This vacuum curvature is expressed in terms of what is called the cosmological constant. The familiar gravitational force between material objects is always attractive. A positive cosmological constant produces a repulsive gravitational force.

At the time, Einstein and most others assumed that the stars formed a fixed, stable firmament, as the biblical phrasing goes. A stable firmament is not possible with attractive forces alone, so Einstein thought that the repulsion provided by the cosmological constant might balance things out. When, soon after, Edwin P. Hubble discovered that the universe was not a stable firmament but expanding, the need for a cosmological constant was eliminated; Einstein called it his "biggest blunder." Until recently, all the data gathered by astronomers have fit very well to models that set the cosmological constant equal to 0.

Einstein's blunder resurfaced in 1980 with the inflationary model of the early big bang, which proposed that the universe underwent a huge exponential expansion during its first 10^{-35} second or so (Kazanas 1980, Guth 1981, Linde 1982). One way to achieve exponential expansion is with the curvature produced by a cosmological constant in otherwise empty space.

This was not all. In 1998, two independent research groups studying distant supernovae were astonished to discover, against all expectations, that the current expansion of the universe is accelerating (Reiss et al. 1998, Perlmutter et al. 1999). The universe is falling up! Once again, gravitational repulsion is indicated, possibly provided by a cosmological constant.

Whatever is producing this repulsion, it represents 70 percent of the total mass-energy of the universe—the single largest component. This component has been dubbed *dark energy* to distinguish it from the gravitationally attractive *dark matter* that constitutes another 26 percent of the mass-energy. Neither one of these ingredients is visible, nor can they be composed of ordinary atomic and subatomic matter like quarks and electrons. Familiar luminous matter, as seen in stars and galaxies, comprises only 0.5 percent of the total mass-energy of the universe, with the remaining 3.5 percent in ordinary but nonluminous matter like planets.

If dark energy is in fact the vacuum energy implied by a cosmological constant, then we have a serious puzzle called the cosmological constant problem (Weinberg 1989). As the universe expands, regions of space expand along with it. A cosmological constant implies a constant energy density, and the total energy inside a given region of space will increase as the volume of that region expands. Since the end of inflation, volumes have expanded by 120 orders of magnitude. This implies that the cosmological constant was fine-tuned to be 120 orders of magnitude below what it is now, a tiny amount of energy. If the vacuum energy had been just a hair greater at the end of inflation, it would be so enormous today that space would be highly curved, and the stars and planets could not exist.

Design advocates have not overlooked the cosmological constant problem (Ross 1998). Once again, they claim to see the hand of God in fine tuning the cosmological constant to ensure that human life, as we know it, can exist. Recent theoretical work, however, has offered a plausible nondivine solution to the cosmological constant problem.

Theoretical physicists have proposed models in which the dark energy is not identified with the energy of curved space-time but with a dynamical, material energy field called quintessence. In these models, the cosmological constant is exactly 0, as suggested by a symmetry principle called supersymmetry. Since 0 multiplied by 10^{120} is still 0, we have no cosmological constant problem in this case. The energy density of quintessence is not constant but evolves along with the other matter-energy fields of the universe. Unlike the cosmological constant, quintessence energy density need not be fine-tuned.

While quintessence may not turn out to provide the correct explanation for the cosmological constant problem, it demonstrates, if nothing else, that science is always hard at work trying to solve its puzzles within a materialistic framework. The assertion that God can be seen by virtue of his acts of cosmological fine tuning, like intelligent design and earlier versions of the argument from design, is nothing more than another variation on the disreputable God-of-the-gaps argument. These rely on the faint hope that scientists will never be able to find a natural explanation for one or more of the puzzles that currently have them scratching their heads and therefore will have to insert God as the explanation. As long as science can provide plausible scenarios for a fully material universe, even if those scenarios cannot be currently tested, they are sufficient to refute the God of the gaps.

An Infinity of Universes

We have shown that conditions that might support some form of life in a random universe are not improbable. Indeed, we can empirically estimate the probability that a universe will have life. We know of one universe, and that universe has life; so the measured probability is 100 percent, albeit with a large statistical uncertainty. This rebuts a myth that has appeared frequently in the design literature and is indicated by Barrow and Tipler's (1986) option 3: that only a multiple-universe scenario can explain the coincidences without a supernatural creator (Swinburne 1998). Multiple universes are certainly a possible explanation; but a multitude of other, different universes is not the sole naturalistic explanation available for the particular structure of our universe.

But if many universes beside our own exist, then the anthropic coincidences are a no-brainer. Within the framework of established knowledge of physics and cosmology, our universe could be one of many in a super-universe, or multiverse. Andrei Linde (1990, 1994) has proposed that a background space-time "foam" empty of matter and radiation will experience local quantum fluctuations in curvature, forming many bubbles of false vacuum that individually inflate into mini-universes with random characteristics. Each universe within the multiverse can have a different set of constants and physical laws. Some might have life in a form different from ours; others might have no life at all or something even more complex or so different that we cannot even imagine it. Obviously we are in one of those universes with life. Other multiverse scenarios have also been discussed (Smith 1990; Smolin 1992, 1997; Tegmark 2003).

Several commentators have argued that a multiverse cosmology violates Occam's razor (entities should not be multiplied beyond necessity) (Ellis 1993). This argument is debatable. The entities that Occam's law of parsimony forbids us from "multiplying beyond necessity" are independent theoretical hypotheses, not universes. For example, the atomic theory of matter multiplied the number of bodies we must consider in solving a thermodynamic problem by 10^{24} or so per gram. But it did not violate Occam's razor. Instead, it provided for a simpler, more powerful, more economic exposition of the rules that were obeyed by thermodynamic systems. The multiverse scenario is more parsimonious than that of a single universe. No known principle rules out the existence of other universes, which furthermore are suggested by modern cosmological models.

Conclusion

The media have reported a new harmonic convergence of science and religion (Begley 1998). This is more a convergence between theologians and devout scientists than a consensus of the scientific community. Those who deeply need to find evidence for design and purpose in the universe now think they have done so. Many say that they see strong hints of purpose in the way in which the physical constants of nature seem to be exquisitely fine-tuned for the evolution and maintenance of life. Although not so specific that they select out human life, various forms of anthropic principles have been suggested as the underlying rationale.

Design advocates argue that the universe seems to have been specifically designed so that intelligent life would form. These claims are essentially a modern, cosmological version of the ancient argument from design for the existence of God. The new version, however, is as deeply flawed as its predecessors were, making many unjustified assumptions and being inconsistent with existing knowledge. One gross and fatal assumption is that only one kind of life, ours, is conceivable in every possible configuration of universes. But a wide variation of the fundamental constants of physics leads to universes that are long-lived enough for life to evolve, even though human life need not exist in such universes.

Although not required to negate the fine-tuning argument, which collapses under its own weight, other universes besides our own are not ruled out by fundamental physics and cosmology. The theory of a multiverse composed of many universes with different laws and physical properties is actually more parsimonious, more consistent with Occam's razor, than a single universe. Specifically, we would need to hypothesize a new principle to rule out all but a single universe. If, indeed, multiple universes exist, then we are simply in that particular universe of all the logically consistent possibilities that had the properties needed to produce us.

The fine-tuning argument and other recent intelligent-design arguments are modern versions of God-of-the-gaps reasoning, in which a God is deemed necessary whenever science has not fully explained some phenomenon. When humans lived in caves, they imagined spirits behind earthquakes, storms, and illness. Today, we have scientific explanations for those events and much more. So those who desire explicit signs of God in science now look deeper, to highly sophisticated puzzles like the cosmological-constant problem. But once again, science continues to progress, and we now have a plausible explanation that does not require fine tuning. Similarly, science may someday have a theory from which the values of existing physical constants can be derived or otherwise explained.

The fine-tuning argument would tell us that the sun radiates light so that we can see where we are going. In fact, the human eye evolved to be sensitive to light from the sun. The universe is not fine-tuned for humanity. Humanity is fine-tuned to the universe.

Chapter 13

Is Intelligent Design Science?

MARK PERAKH AND MATT YOUNG

T HE PRECEDING CHAPTERS of this book have attacked the scientific basis for intelligent-design creationism. Some readers may therefore infer that our dispute with the advocates of intelligent design is purely scientific; that intelligent design is a legitimate scientific theory on a par, say, with evolution; and that it is just an alternate way of attacking the problem of origins. Indeed, we risk legitimizing intelligent design simply by engaging it.

Let us make clear, then, that we do not consider intelligent design to be a legitimate scientific endeavor. Intelligent design is not bad science like cold fusion or wrong science like the Lamarckian inheritance of acquired characteristics, although it probably lies farther along the same continuum. Criticizing it gives it no more scientific legitimacy than criticizing astrology—no more than the magazine *Skeptical Inquirer* gives to quack medicine when it exposes such practices as phony.

Looking for the footprints of the deity is not necessarily unscientific. What is unscientific is to decide ahead of time on the answer and search for God with the determination to come up with a positive result. That is precisely what William Dembski, Michael Behe, and other ID advocates seem to be attempting. Knowing the answer in advance and being immune to contradictory evidence are typical of pseudoscience.

Perhaps we should be hesitant to use a label such as pseudoscience or crank science; after all, such terms are no longer favored among philosophers of science. It has become increasingly clear (Laudan 1988) that there is no clean way of separating scientific claims from nonscientific just by applying principles like falsifiability or methodological naturalism. Additionally, labeling a rival idea as pseudoscientific may well replace real argument with a political attempt to deny it legitimacy.

Nevertheless, we argue that pseudoscience can be a useful term. If the intelligent-design advocates advertise themselves as doing science, even when their practices are far from the customary intellectual conduct of mainstream science, we can and should suspect that intelligent design is not legitimately science. This suspicion is not a substitute for the detailed scientific critiques offered in the preceding chapters. Nevertheless, exploring whether the label *pseudoscience* applies may help us better understand what is wrong with intelligent design.

Before rendering judgment on intelligent design, however, let us examine some pseudosciences and see what they have in common and why we call them pseudosciences.

Some Features of Pseudoscience

DENIAL OF ESTABLISHED SCIENTIFIC FACT

Homeopathy provides a good example of a pseudoscience that denies known facts. Specifically, homeopathic "physicians" start with a chemical compound that is thought to cure a disease, dissolve that compound in water, and then repeatedly dilute it many times until there are, at most, just a few molecules of the original compound in the solution. Indeed, the water contains impurities in many times the concentration of the "medication."

Homeopaths are aware of the dilution problem and rely on an ad hoc hypothesis: the water remembers what has been put into it. How? Jacques Benveniste, a French medical doctor, says vaguely that some sort of electromagnetic radiation stays in the water (Lawren 1992, Friedlander 1995). He has, however, not measured this radiation and has apparently forgotten that electromagnetic radiation travels at the speed of light and would be gone from his solution in a few nanoseconds, at most.

Young-earth creationists claim that the earth is approximately 10,000 years old. When presented with fossil evidence to the contrary, some propose that God put the fossils into the earth for a reason that we do not know. By proposing such an ad hoc hypothesis, they make it impossible to measure the age of the earth: it is 10,000 years by fiat.

In the same way, former astronomer Hugh Ross (1998) accepts the Hebrew Bible's claim that people lived for 900 years around the time of Noah and postulates that God created the Vela supernova specifically to bathe the earth in cosmic rays, cause genetically programmed cell death, and shorten our life spans to a mere 120 years. Ross's hypothesis—for that is all it is—would be more convincing if cosmic rays accounted for most of the radiation on our gonads, but they do not (Young 2000). Background radiation due to radioactive minerals in the environment contributes more than half that radiation. Even before the Vela supernova, there was plenty of radiation to initiate programmed cell death.

What are regarded as established facts may turn out to be wrong. But we need very strong reasons to suspect a major mistake. Ad hoc scenarios such as Beneviste's are not enough.

UNTESTABLE HYPOTHESES

Invoking an ad hoc hypothesis to explain a result you did not expect is not necessarily bad science. When a certain nuclear disintegration did not appear to obey the law of conservation of energy, the physicist Wolfgang Pauli postulated the existence of a new particle—the neutrino. Pauli's postulate was not the end of the argument, however; it was the beginning. He calculated the properties of the neutrino and thereby allowed scientists to search for such a particle. A particle with the required properties was found approximately two decades later. If it not been found or had had the wrong properties, scientists would have had to seek another solution to the problem.

Pauli's hypothesis gave a plausible explanation of why the energy seemed not to be conserved. But more important, it was very specific and could be tested. Contrast that with Benveniste's hypothesis, which gives no idea of the properties of the radiation, how it got into the water, why it stays around, or how to detect it.

A hypothesis has to be testable, or else it is useless as a scientific tool. Both confirming and disconfirming evidence weigh in, although neither can be wholly conclusive (Bunge 1996, 180–83). Nevertheless, a good test has to risk failure, and a good scientist recognizes failure.

A theory that explains everything also explains nothing, because it cannot be tested. For example, an astrologer might say that a person who is born with Mars in his house should be aggressive. If the astrologer sees a submissive person who was also born with Mars in his house, he says, "That sometimes happens; he felt the strength of Mars inside him, could not handle it, and retreated into himself" (Dean 1986–87). In other words, a person born with Mars in his house will be either aggressive or submissive. Astrologers who always manufacture excuses to protect their theories from failure are not practicing science (Perakh 2002b).

TRIES TO "PROVE THAT"

A pseudoscientist tries to prove *that* something is true; a good scientist tries to find out *whether* it is true. This distinction is important. If we attack a problem, certain of the answer, then we will find that answer, whether it is right or not. Benveniste found a positive result when he performed certain tests; but when his procedure was tightened up by an international visiting committee, the positive result disappeared.

William Dembski (1998d) forfeits his credibility as a scientist, or ought to, when he says, "As Christians, we know" (14). Sorry, but we don't know. What Dembski ought to say is "As Christians, we hypothesize," and then go out and test his hypothesis. Instead, he seems to have the answer and therefore only pretends to be searching for it. The fact that others, non-Christians, claim to have different answers ought to give Dembski pause, but it apparently does not.

EVERYONE IS WRONG BUT US

Pseudoscientists seem to think that everyone is wrong but them. Indeed, they may dare you to prove them wrong, little realizing that the burden of proof is usually on the person who makes the claim. Often they imply a conspiracy among their opponents to silence them. Many pseudoscientists make grandiose claims and think they are misunderstood geniuses; they compare themselves to Galileo, a man persecuted by the Church for his scientific discoveries (Friedlander 1995). Behe (1996) claims that his thesis of irreducible complexity "must be ranked as one of the greatest achievements in the history of science. The discovery rivals those of Newton and Einstein, Lavoisier and Schrödinger, Pasteur, and Darwin" (233). Dembski (1991), at the beginning of his career, had already compared himself to Kant and Copernicus and now claims (1999, 2002b) he has discovered a new law of thermodynamics. The philosopher Rob Koons compares Dembski to Newton (Dembski 1999, jacket blurb).

Why is the burden of proof on the claimant? Largely for practical reasons. Most scientists have no time to evaluate every unsupported claim that passes their way. They may miss some important ideas, like continental drift, but they will more likely miss a lot more bad science and pseudoscience. If you want to get scientists' attention, you have to provide something concrete, supported by evidence: something they can evaluate rigorously.

Additionally, it is often hard to prove the negative of a statement. Until humans landed on the moon, we could not have disproved the old maxim, "The moon is made of green cheese." Instead, we had to ask the proponents of that theory to provide evidence of their claim. Since they could not, we did not accept it. But neither did we make an all-out effort to prove that the moon was not made of green cheese. It was simply not worth our while, and we would have been hard-pressed to find an argument that would have convinced the true believers.

OTHER FEATURES OF PSEUDOSCIENCE

Distinguishing pseudoscience from wrong science is not always cut and dried, although an honest person who practices wrong science will usually admit error when error is proved. Soviet scientist Boris Deryagin, for example, thought that he had discovered a polymerized form of water, which he called polywater (Levi 1973, Friedlander 1995). Many others thought that they had detected polywater, too, before evidence against its existence began to accumulate. When the evidence proved that the polywater was, in essence, a solution of glass (silicon dioxide) in water, Deryagin conceded his error. A pseudoscientist would not have done so; the pseudoscientist rarely, if ever, admits error but finds some way to patch up the theory. This is not to say that science is always right. Rather, wrong science is correctable, whereas pseudoscience is not.

Finally, pseudoscientists often use made-up terms and vague concepts that hide their lack of intellectual substance.

Methodological Naturalism

Advocates of intelligent design respond by claiming that "official" science unjustifiably views as legitimate objects of its inquiry only whatever is natural and rejects out of hand everything supernatural, leaving no place for intelligent design (Johnson 1993, 1998; Behe 1996; Dembski 1999; Wells 2000). Such an attitude supposedly stems from the dogmatic philosophical presuppositions of the scientific establishment and is claimed to be an obstacle to free inquiry.

Methodological naturalism has indeed been a feature of science, but only as a practical matter and not as a fundamental principle. Methodological naturalism has so far worked and enabled science to achieve great success. In fact, however, science differentiates only between known and unknown, or between explained and unexplained, not between natural and supernatural. Every phenomenon that can be studied using methods of inquiry based on evidence is legitimate in science.

As an example of how science can legitimately approach a problem regardless of its possible supernatural implications, let us consider the affair of the Bible codes.

The Bible Codes

In 1994, the peer-reviewed journal *Statistical Science* printed an article by Doron Witztum, Eliyahu Rips, and Yoav Rosenberg (WRR) claiming discovery of a meaningful code hidden in the Hebrew text of Genesis. WRR defined an *equidistant letter sequence* (ELS) as a meaningful word that can be formed in a text by sequentially extracting letters separated by equal intervals, or skips. For example, look at the title of this section. Ignoring spaces, we write it as THEBIBLECODES. The first, fifth, and ninth letters of that string form an ELS (with a skip of 4) for the word TIC. WRR noted that Genesis contains a large number of ELS's. Their claim is true. The same, though, is equally true of any sufficiently long text in any language. With a suitable computer program, thousands of ELS's with various skips can be instantly identified in every text—in the Manhattan phone book as well as in the Bible.

Although WRR did not mention intelligent design, the problem they faced was very similar to examples discussed by ID advocates (Behe 1996; Dembski 1998a, 1998d, 1999, 2002b): they wanted to determine whether the ELS's in Genesis could be attributed to chance or whether design had to be inferred (see chapter 9 in this book).

WRR conducted a computerized statistical experiment. They compiled a list of famous rabbis who lived between early medieval times and the eighteenth century. Their computer program located ELS's that spell the appellations of those rabbis, with various skips, as well as ELS's that spell the dates of birth and/or death of the same rabbis. (In Hebrew, dates are expressed by letters of the alphabet.) They estimated the statistically averaged distance within the text between the ELS's for the appellations and for the dates of birth and/or death of the same rabbis. Then their program created one million permuted lists of appellations and dates; the appellations for individual rabbis and their dates became mismatched in these permuted lists.

WRR calculated the statistically averaged distance between ELS's for appellations and dates for all the permuted lists and compared those distances with the distances in the original list. They concluded that the ELS's for the appellations and for the dates of the same rabbi in the text of Genesis are situated statistically much closer to each other than the distances between ELS's for appellations and dates if found for different rabbis. They estimated that the probability of such an "unusually close proximity" happening by chance did not exceed 1 in 62,000.

Since the rabbis in question all lived much later than Genesis was written, the unusual proximity of the encoded rabbis' names to their encoded dates of birth or death means the text's author must have known the future. In other words, WRR's article alleged scientific proof of a miracle.

If we believe the ID advocates, the scientific establishment, represented by the editorial board of *Statistical Science*, should have rejected WRR's paper out of hand because it dealt with the supernatural. On the contrary, they published WRR's paper, although the referees had expressed serious doubts about WRR's statistical procedure.

A number of experts analyzed WRR's procedure and found that it suffered from a number of irregularities. Gradually, specialists in statistics and related fields came to an overwhelming consensus that WRR's data were unreliable. *Statistical Science* published a paper that decisively showed WRR's methodology to be contrary to the requirements of scientific rigor; hence, their results could not be trusted (McKay et al. 1999). Additional critiques of WRR's work appeared elsewhere (Simon 1998; Perakh 2004b; Hasofer 1998; Ingermanson 1999; Cohen 2000).

The results claimed by WRR were rejected not because the object of their study violated methodological naturalism but because of the faults in their procedure. If WRR's data had been statistically sound, then there would have been sufficient reason to consider a nonchance origin of the code in Genesis. Further, once chance is dismissed as the cause of the "close proximity," all sorts of alternative explanations become legitimate alternatives for a scientific discourse. Among possible alternative inferences, for example, are time travel, psychic prediction of the future, and extraterrestrials as the authors of Genesis (Raël 1986) as well as, yes, inferring the existence of a disembodied intelligent designer.

If WRR's data were statistically sound, scientists would include the inference to intelligent design as one among many possibly legitimate explanations. No naturalistic philosophical predispositions would prevent the inference to the supernatural. Scientists rejected such an inference only because WRR's data were found unsatisfactory, for both statistical (McKay et al. 1999) and extra-statistical (Perakh 2004b) reasons. An inference to the supernatural has not yet been accepted in any other case either—again only because no *evidence* for such superhuman intelligent design has been demonstrated. Until such evidence is unearthed, the supernatural will not become a part of genuine science.

Science is neither based on methodological naturalism nor restrained by it, and likewise it is not restrained by any other metaphysical principle. It is restrained by one and only one requirement: it requires evidence.

Genuine Science versus Dembski's Z-Factors

Regarding the question of whether intelligent design is genuine science or pseudoscience, it also seems relevant to review the concept that Dembski

(2002b) calls Z-factors. He defines these as "some entity, process or stuff out-side the known universe [which] . . . purports to solve some problem of gen-eral interest and importance" (87). Dembski defines the *inflationary fallacy* as estimating the probability of an event based not only on reliable knowledge but also on an arbitrary additional hypothesis, which he calls a Z-factor. A Z-factor is an unjustified excursion beyond the available knowledge (in Dembski's terms *inflation of probabilistic resources*), which can be used to estimate the prob-ability of an event.

Dembski discusses four Z-factors: the bubble universes of Alan Guth's (1997) inflationary cosmology, the many worlds of Hugh Everett's (1957) in-terpretation of quantum mechanics, the self-reproducing black holes of Lee Smolin's (1997) cosmological natural selection, and the possible worlds of David Lewis's (1986) extreme modal realist metaphysics. These four concepts postulate the existence of many (so far undetected) universes besides our own. They offer different assumptions regarding the origin of the multiple universes and their putative properties. None has a direct empirical basis, but their au-thors have proposed arguments in favor of their plausibility, and each of these hypotheses has a certain explanatory potential regarding the structure and the history of our universe.

The available knowledge about the structure and history of the universe is insufficient to choose among the hypotheses by Guth, Smolin, Everett, and Lewis. Dembski therefore calls all four concepts Z-factors. The four Z-factors in question are indeed speculative, but that is not the issue in this chapter. We are, rather, interested in some of the arguments Dembski suggests against the inflationary fallacy.

In Dembski's (2002b) view, "Each of the four Z-factors considered here possesses explanatory power in the sense that each explains certain relevant data and thereby solves some problem of general interest and importance" (90). He continues, however, to say that possessing explanatory power is not suffi-cient for accepting a theory. What is also necessary is independent evidence in favor of that theory: "Independent evidence is by definition evidence that helps establish a claim apart from any appeal to the claim's explanatory power. . . . It is a necessary constraint on theory construction so that theory construction does not degenerate into total free-play of the mind" (90).

As an example, Dembski discusses a hypothetical gnome theory of fric-tion: "suitably formulated, the gnome theory of friction can explain how ob-jects move across surfaces just as accurately as current physical theories. So, why do we not take the gnome theory of friction seriously? One reason . . . is the absence of independent evidence for gnomes" (91).

Even if we disregard Dembski's dubious assertion that the gnome theory can explain friction as well as current physical theories, we can agree with him that a plausible theory has to offer explanatory power (otherwise, it is not useful) and be supported by independent evidence (otherwise, it would "degenerate into total free-play of the mind" [90]). Do intelligent-design theories—in particular, Dembski's—provide explanatory power? Are they supported by independent evidence? We suggest that the answers are unequivocally no.

ID theory claims that we can establish design by some rational procedure, whose principal features are encapsulated in Dembski's explanatory filter. Detailed analyses of the explanatory filter may be found in chapter 8 of this book and elsewhere (Chiprout 2003; Elsberry 1999, 2000; Fitelson et al. 1999; Perakh 2001b, 2002a, 2004a; Wilkins and Elsberry 2001; Elsberry and Shallit 2002). Here, we are interested only in answering the questions about explanatory power and independent evidence.

Does intelligent-design theory provide explanatory power? If so, it must provide information about the details of the design and, to this end, about the nature of the designer. ID theory, however, deliberately avoids the answers to this question. Advocates of intelligent design (Dembski 1999) insist that their theory is not tied to any concept of a designer but just provides a means to distinguish among chance, regularity, and design as the causal antecedents of the event in question.

The designer in the intelligent-design theory looks like another Z-factor. Indeed, Dembski's concept is based on a much more egregious inflationary fallacy than those of his four offenders. Guth (1997) and Smolin (1997) at least suggest ideas in regard to the features, properties, and behavior of their Z-factors. By contrast, Dembski deliberately leaves beyond consideration the attributes of the Z-factor in his own theory. Moreover, Guth and Smolin have suggested certain ideas for indirect tests of their Z-factors, while the advocates of intelligent design propose nothing even close to an empirical test.

In fact, the intelligent-design theory does not have explanatory power. To simply state that an event is due to intelligent design explains nothing because the term *designer* has no defined meaning in the theory and the modes of the designer's activities remain mysterious and unexplained. Indeed, both Behe (1996) and Dembski (1998a, 2002b) refuse to even speculate on the attributes of the designer, whose existence can supposedly be asserted using the design inference (Dembski 1998a, 1999, 2002b). Hence, while Dembski has stated the necessity of explanatory power for any useful theory, he forgets about that requirement when turning to his own theory.

Advocates of intelligent design usually refuse to discuss the nature of the alleged designer. They try to deflect criticism of their refusal by citing examples in which design is inferred despite the lack of knowledge about the designer. For example, Dembski asks, Is the design inference legitimate in the case of Stonehenge? We all agree that it is. He says, however, that nothing is known about the designer in that case either. On the contrary, as regards Stonehenge and similar cases, we infer a well-known type of a designer: a human designer. We attribute the creation of Stonehenge to a human designer precisely because we know so much about the features of human design and see those features in the object observed.

In another example, Dembski (2002b) refers to a book by Del Ratzsch (2001). Ratzsch suggests imagining that an expedition to some planet of the solar system finds a bulldozer standing in a field. Obviously, we conclude that the bulldozer was designed rather than that it happened to exist by sheer chance. Dembski insists, however, that we infer a designer without any knowledge about the designer and his characteristics. Why cannot the same attitude be applied to a mysterious designer in his design theory?

The fallacy of such an argument is evident. A bulldozer on an alien planet has features that testify not only to certain characteristics of its supposed designer but also to the possible use for which it was designed. The bulldozer has treads evidently designed for motion, a seat evidently designed to accommodate a creature anatomically similar to earthly humans, pedals evidently designed for feet similar to those of earthly humans, and many other features providing good ideas about what kind of a designer must be responsible for the observed object and what use it was intended for. Additionally, we have prior experience of bulldozers and know that they are artifacts.

In Ratzsch's terms, a bulldozer displays an obvious *artifactuality*. In Niall Shanks and Karl Joplin's (1999) terms, the bulldozer is "antecedently recognizable as an artifact" (269). It is precisely because we know so much about both bulldozers and the humans who design them that we would infer design if a bulldozer were found on Mars. Moreover, the design inference in the case of that bulldozer will be made without any reference to Dembski's explanatory filter, which is utterly useless for inferring design in the hypothetical case under discussion.

The bulldozer example is irrelevant for many other reasons as well. A bulldozer is not a living organism that can reproduce, develop, or evolve spontaneously. Indeed, the design inference is controversial only with respect to biological entities. When we see a bulldozer or a poem, there is no controversy; we unequivocally attribute them to design because of our extensive knowledge about such objects and the human designers who create them. In contrast to bulldozers and poems, organisms are not designed but inherit genes from their forebears. A bulldozer is designed by an engineer and built by laborers according to the engineer's design. Dembski and his colleagues seem peculiarly blind to the obvious difference between artifacts and organisms.

We know nothing whatsoever about the alleged disembodied designer of the intelligent-design theory or about what that designer's creations should look like. The case is therefore very different from bulldozers and poems. A reference to such a designer lacks explanatory power.

According to Dembski, an even more important requirement for a theory is independent evidence supporting that theory. Strangely, he and his colleagues in the intelligent-design enterprise forget about the criterion of independent evidence as soon as they turn to their own theories. Where is independent evidence supporting the intelligent-design theory? There is none. The absence of explanatory power and of independent evidence, according to Dembski's own criteria, signifies the degeneration of the theory into a "total free-play of the mind," which Dembski seems to disapprove of for all theories except his own. Despite their substantial financial resources, the advocates of intelligent design have so far failed to come up with a real scientific research program and indulge instead in philosophical or theological discourses and political maneuvering (Forrest 2001).

Dembski views explanatory power as a category independent of evidence. Indeed, the gnome theory of friction, in his view, can provide explanatory power despite lack of evidence for the existence of gnomes. Explanatory power without evidence is, however, meaningless. The gnome theory of friction does not plausibly explain friction precisely because there is no evidence of the gnomes' existence. The gnome theory is pure speculation and has no explanatory power. Explanatory power is meaningful only if it is based on facts and evidence. To say that friction is caused by gnomes is to explain nothing because we have no knowledge about what kind of a behavior those postulated gnomes may have. In fact, a theory has plausible explanatory power only if it is also supported by evidence.

Dembski's intelligent designer is just another Z-factor. His separation of explanatory power from evidence is contrived; it is no more than the "free-play of the mind," a mote that he finds in the eyes of others.

But Is It Pseudoscience?

To decide whether intelligent design is science, we have to first realize that the existence or nonexistence of God is a fact. Whether we can find out anything about this fact is problematic, but if we are going to do so, we have to deal with facts and evidence. Only an objective evaluation based on evidence is apt to find God's putative footprints in the natural world. Faith, in this context, is a blind alley because it stifles rigorous investigation; indeed, if we accept Michael Behe's concept of irreducible complexity, we might as well throw in the towel and not even try to understand the evolution of the flagellum (see chapter 6 in this book).

Unlike young-earth creationists, the intelligent-design neocreationists do not usually deny scientific fact; rather, they work their theories around or into what is already known. Their hypothesis—that God might have left behind evidence of his creation—is not indefensible, unless we are willing to rule out theism from the beginning. As we have seen, however, scientists' commitment to methodological naturalism does not mean they need such an a priori assumption.

On the other hand, although they are sometimes coy about the identity of their intelligent agent, the neocreationists plainly try to prove *that*, not find out *whether*—a clear feature of pseudoscience. Additionally, they say everyone is wrong but them and compare themselves with, say, Copernicus, Newton, and Boltzmann. They often imply a conspiracy among their opponents to exclude religion or God from science, another common feature of pseudoscience.

Of the criteria we consider relevant to the issue, the advocates of intelligent design pass on only two: they do not usually deny known facts (although sometimes they do), and their hypothesis that the universe may have been designed is not indefensible. Where it counts—assuming the answer, implying conspiracy, inflating the importance of their alleged breakthroughs, and lacking evidence—their work has enough features to be recognized as pseudoscience.

Intelligent design is the argument from design in new clothing. Advocates of intelligent design, such as Dembski, claim to look for evidence of design, but they simply estimate probabilities and use them to eliminate chance or necessity. They ignore other alternatives and have no positive criterion for identifying a designer; their combination of low probability (often miscalculated) with a dubiously defined and often misused concept of specification provides no real evidence.

It is not scientific.

Acknowledgment

Taner Edis contributed valuable insights to this chapter.

Appendix

Organizations and Web Sites

COMPILED BY GARY S. HURD

A GREAT DEAL of the basic content of the intelligent-design movement is published on the Internet. Although we have included many web links in the reference list, we want to expand the list of sources, both pro and con, as a help to our readers. Bear in mind, however, that estimates of the longevity of web addresses suggest that many of these sites may soon become unavailable. Commercial search engines such as Google (http://www.google.com/search) and Yahoo (http://www.yahoo.com) are valuable sources of up-to-date information.

In the following list, all quotations appear on the web sites in question, unless otherwise noted.

Sources That Support Intelligent-Design Creationism

Origins.org (http://www.origins.org/menus/design.html)
    "Origins.org focuses primarily on the scientific theory known as Intelligent Design and reaches one logical conclusion: that the universe and life show verifiable signs of intelligent creation because there is an intelligent Creator. Some of our resources deal with scientific data exclusively and some take the defensible position that the data point to and support the Biblical claim of Divine Creation. We let the resources speak on their own merits." Contributors

include Jonathan Wells, Charles B. Thaxton, Hugh Ross, Phillip E. Johnson, William A. Dembski, Paul Davies, and Walter L. Bradley.

Intelligent Design and Evolution Awareness (IDEA) Club
(http://www.ucsd.edu/~idea)

"The Intelligent Design and Evolution Awareness (IDEA) Club birthed in May of 1999 after UC Berkeley law professor Phillip Johnson came and lectured at UCSD. Known for his books critiquing Darwinian evolution, naturalistic thought, and his leadership in the 'Intelligent Design movement,' Johnson was brought to UCSD by UCSD Intervarsity Christian Fellowship and Campus Crusade for Christ to speak on issues related to creation and evolution."

Access Research Network (http://www.arn.org/)

"We focus on such controversial topics as genetic engineering, euthanasia, computer technology, environmental issues, creation/evolution, fetal tissue research, AIDS, and so on."

Intelligent Design Undergraduate Research Center
(http://www.idurc.org/)

This web site was begun by a University of California, San Diego, student and has become a source of much information about ID. It is a division of Access Research Network.

The International Society for Complexity, Information, and Design (ISCID)
(http://www.iscid.org/)

"[This] cross-disciplinary professional society . . . investigates complex systems apart from external programmatic constraints like materialism, naturalism, or reductionism. The society provides a forum for formulating, testing, and disseminating research on complex systems through critique, peer review, and publication. Its aim is to pursue the theoretical development, empirical application, and philosophical implications of information- and design-theoretic concepts for complex systems."

Design Inference Web Site: The Writings of William A. Dembski
(http://www.designinference.com)

"A mathematician and a philosopher, William A. Dembski is associate research professor in the conceptual foundations of science at Baylor University and a senior fellow with Discovery Institute's Center for Science and Culture in Seattle. He is also the executive director of the International Society for Complexity, Information, and Design."

Discovery Institute—Center for Science & Culture
(http://www.discovery.org/csc)

Formerly the Center for the Renewal of Science & Culture. The Discovery Institute is the home base for most of the intelligent-design creationism movement.

Evolution vs. Design: Is the Universe a Cosmic Accident or Does It Display Intelligent Design? (http://www.evidence.info/design/)

This site, hosted by Evidence for God.info, is a good example of the success that the intelligent-design creationists have had glossing over basic differences between what they say and what they mean.

Intelligent Design Network, Inc. (http://www.intelligentdesignnetwork.org)

"Intelligent Design Network, Inc. is a member based nonprofit organization. Idnet was organized in 1999 in connection with the debate over the Kansas Science Education Standards.

"Intelligent Design is a scientific theory that intelligent causes are responsible for the origin of the universe and of life and its diversity. It holds that design is empirically detectable in nature, and particularly in living systems.

"Intelligent Design is an intellectual movement that includes a scientific research program for investigating intelligent causes and that challenges naturalistic explanations of origins which currently drive science education and research."

Can Intelligent Design (ID) Be a Testable, Scientific Theory?
Evolution vs. Design: Is the Universe a Cosmic Accident or Does it Display Intelligent Design? (http://www.godandscience.org/evolution/)

These web pages, hosted by God and Science.org, illustrate the intelligent-design support among young-earth creationists.

Organizations That Support Science

Far fewer web sites are devoted exclusively to the creationism-versus-science issue and advocate science per se. Much legitimate scientific information on the Internet, however, supports basic science and therefore opposes intelligent-design creationism. See, for example, Wilkinson Microwave Anisotropy

Probe—Cosmology (http://map.gsfc.nasa.gov/index.html) and Kimball's Biology Pages (http://users.rcn.com/jkimball.ma.ultranet/BiologyPages). We first list web sites that are directly involved in opposing creationist efforts to insert ID into public school curriculums. These include both professional and volunteer organizations.

SOURCES THAT OPPOSE CREATIONISM

National Center for Science Education (NCSE) (http://www.ncseweb.org)

"We are a nationally-recognized clearinghouse for information and advice to keep evolution in the science classroom and 'scientific creationism' out. While there are organizations that oppose 'scientific creationism' as part of their general goals (such as good science education, or separation of church and state), NCSE is the only national organization that specializes in this issue. When teachers, parents, school boards, the press and others need information and help, they turn to NCSE."

AntiEvolution.org (http://www.antievolution.org/)

This web site is for the critical examination of the anti-evolution movement. Unlike anti-evolution–advocacy web sites, this site aims to provide links to both the anti-evolutionists making their own arguments and also to the critics who provide mainstream-scientific answers to those arguments. The site's discussion forum can yield particularly useful observations and references.

Talk Reason (http://www.talkreason.org/)

"This website presents a collection of articles which aim to defend genuine science from numerous attempts by the new crop of creationists to replace it with theistic pseudo-science under various disguises and names. Talk Reason is designed to provide a forum for articles arguing against modern creationism in all of its forms."

Talk Origins (http://www.talkorigins.org)

Talk Origins is the grandmother of anti-creationist web sites. "The Talk.Origins archive is a collection of articles and essays most of which have appeared in talk.origins [news group] at one time or another. The primary reason for this archive's existence is to provide mainstream scientific responses to the frequently asked questions (FAQs) that appear in the talk.origins newsgroup and the frequently rebutted assertions of those advocating Intelligent Design or other creationist pseudosciences."

Talk Design (http://www.talkdesign.org)
"This web site, a sub-site of TalkOrigins.org, is a response to the 'Intelligent Design' movement of creationism. It is dedicated to:

"Assessing the claims of the Intelligent Design movement from the perspective of mainstream science

"Providing an archive of materials that critically examine the scientific claims of the ID movement.

Was Darwin Wrong (http://www.wasdarwinwrong.com)
A collection of thoughtful and thorough book reviews by Gert Korthof emphasizing works by critics of evolutionary biology.

SCIENCE EDUCATION AND PROFESSIONAL SCIENCE ORGANIZATIONS
American Institute of Biological Sciences (http://www.aibs.org/outreach/evlist.html)
"In 1947, the American Institute of Biological Sciences was federally chartered as a non-profit scientific organization to advance research and education in the biological sciences.

"The AIBS/NCSE Evolution List Server Network, for the U.S. and Canada. Allows scientists, teachers, and other interested parties to be in touch with each other locally, nationally, and internationally. They can facilitate support groups for teachers trying to teach evolution in a difficult atmosphere. They can also permit rapid communications and grass-roots activity when school boards or legislatures are considering policies that will promote the teaching of anti-evolutionary ideas in science classes."

EvoNet: A Worldwide Network for Evolutionary Biology (http://www.evonet.org)
"This website provides the evolution biology community with a variety of resources. These include, but aren't limited to, the areas of research, education, and public outreach. Researchers are provided with listings of people, institutions, software, and websites segregated by areas of interest. Educators can take advantage of listings of classroom resources for institutions of all levels, as well as software and other materials. Public outreach intends to breach communication gaps between our science and interested parties (as well as the general public)."

National Association of Biology Teachers (http://www.nabt.org)
The association publishes *American Biology Teacher*, available to members. There are regular articles on teaching evolution and the political and pedagogic threats to science teaching posed by ID and other forms of creationism.

National Science Teachers Association (http://www.nsta.org)

"The National Science Teachers Association (NSTA), founded in 1944 and headquartered in Arlington, Virginia, is the largest organization in the world committed to promoting excellence and innovation in science teaching and learning for all." Like the National Association of Biology Teachers, the NSTA is a professional organization, and the majority of its web site's content is available only to members.

Biological Sciences Curriculum Study (BSCS) (http://www.bscs.org)

BSCS is a nonprofit organization that develops science curriculums for all grade levels. It was established in 1958 by a grant from the National Science Foundation as one of several new curriculum study groups concerned with improving science education. It will take some searching to find information on this web site.

SINGLE-TOPIC WEB ARTICLES

AAAS Board Resolution Opposing "Intelligent Design" Theory in U.S. Science Classes: November 2002 (http://www.aaas.org/news/releases/2002/1106idIntro.shtml)

"The Board of the American Association for the Advancement of Science has passed a resolution urging policymakers to oppose teaching 'Intelligent Design Theory' within science classrooms, but rather, to keep it separate, in the same way that creationism and other religious teachings are currently handled."

Evolution, Science, and Society (http://evonet.sdsc.edu/evoscisociety)

This is a very attractive web site sponsored by a consortium of science education and professional science organizations. There are two forms of the text: one is based on reports' executive summaries; the other includes expanded treatments of most topics. Some of the expanded texts are hard to find. Try clicking on section titles and headings. Not as pretty, but much easier to read is the text format of the reports (http://www.rci.rutgers.edu/%7Eecolevol/evolution.html).

American Geological Institute (http://www.agiweb.org)

Use the search function from the main page to find publicly available articles such as the following:

"State Challenges to the Teaching of Evolution (5–16–03)" (http://www.agiweb.org/gap/legis108/evolution.html)

"Evolution and the Fossil Record, 2001," by John Pojeta, Jr., and Dale A. Springer (http://www.agiweb.org/news/evolution/)
"AGI Earth Science Education Resources" (http://www.agiweb.org/education/resources.html)

The Evolution Project (http://www.pbs.org/evolution)
This is the web site of the PBS television series, along with teaching resources. A very attractive and easy-to-use site.

CITIZEN GROUPS

Several citizen activist groups formed when ID activists and other creationists launched attacks on local or state educational curriculums. Some of those with web pages are listed here.

Alabama Citizens for Science Education (http://alscience.org/)
"Alabama Citizens for Science Education is a not-for-profit organization which supports the teaching of quality science education in Alabama."

Burlington-Edison (Washington) Committee for Science Education (http://www.scienceormyth.org/)
"BECSE was founded to stop and prevent the teaching of religious doctrine at our high school, and specifically to stop the corruption of the high school's science curriculum by fundamentalist agendas."

Coalition for Excellence in Science and Math Education (CESE) (http://www.cesame-nm.org/)
CESE seeks to "improve the quality and accuracy of the New Mexico Science Content Standards, Benchmarks, and Performance Standards" and "remove the unscientific influence from the state science standards."

Colorado Citizens for Science (CCFS) (http://www.coloradocfs.org/)
"CCFS was organized to counter the political movement of creationism that is attempting to erode quality science education."

Georgia Citizens for Integrity in Science Education (GCISE) (http://www.georgiascience.org/)
"GCISE is dedicated to promoting scientific literacy and excellence in science education. Of primary concern now are attempts to dilute the teaching of evolution and teach so-called 'intelligent design,' an updated version of creationism."

Kansas Citizens for Science (http://www.kcfs.org/)

"Kansas Citizens for Science is a not-for-profit educational organization that promotes a better understanding of what science is, and does, by: 1) advocating for science education, 2) educating the public about the nature and value of science, 3) serving as an information resource."

Michigan Citizens for Science (http://www.michigancitizensforscience.org/)

"Although the theory of evolution is one of the central unifying ideas in science and also one of the most evidentially well-supported of all scientific discoveries, it is continually under attack from those who seek to either eliminate it entirely, or water it down by bringing in religious alternatives dressed up in scientific-sounding language."

Nebraska Religious Coalition for Science Education (http://puffin.creighton.edu/ NRCSE/)

"We oppose the teaching of ID as science, though it might be a subject for study in contexts other than science. It is the NRCSE's position: 1) that evolution is a viable scientific theory, 2) that a Creator is a viable theological proposition, and 3) that creationism ('creation science') and intelligent design theory lack evidence and represent erroneous deviations from the scientific method."

New Mexicans for Science and Reason (http://www.nmsr.org/)

"New Mexicans for Science and Reason (NMSR) consists of individuals, including scientists and non-professionals alike, who share the goals of promoting genuine science, the scientific method, and rational and critical thinking."

Ohio Citizens for Science (OCS) (http://www.ohioscience.org/)

"OCS is a non-profit educational organization committed to improving science literacy in Ohio by bringing Ohio's students into contact with . . . real science and working scientists. We support the teaching of leading scientific theories and methods, . . . including biology, where evolutionary theory is the foundation." The organization publishes *Evolutionary Intelligence: The Monthly Journal of Ohio Citizens for Science*.

Oklahomans for Excellence in Science Education (OESE) (http://www. biosurvey.ou.edu/oese/)

"OESE is a non-profit educational organization that promotes the education of the public about the methods and values of science and advocates excellence in the science curriculum."

Texas Citizens for Science (http://texscience.org/)

"Texas Citizens for Science is a statewide, grassroots organization dedicated to maintaining the professionalism of science education in Texas public schools, the integrity of science in the Texas public school curriculum, and the accuracy of science in Texas government agencies and institutions."

ORGANIZATIONS DEVOTED TO SCIENCE AND RELIGION

Institute on Religion in an Age of Science (IRAS) (http://www.iras.org)

"[The institute is] a non-denominational, independent society with three purposes: 1. to promote creative efforts leading to the formulation, in the light of contemporary knowledge, of effective doctrines and practices for human welfare; 2. to formulate dynamic and positive relationships between the concepts developed by science and the goals and hopes of humanity expressed through religion; 3. to state human values and contemporary knowledge in such universal and valid terms that they may be understood by all peoples, whatever their cultural background and experience, and provide a basis for world-wide cooperation.

"IRAS membership is open to those who have an interest in religion, philosophy, and the natural and social sciences."

IRAS copublishes *Zygon: Journal of Religion and Science*, which prints papers concerning research on religion and science. *Zygon* is published by Blackwell (http://subscrip@blackwellpub.com), and articles are posted at the web site of the Ingenta Institute (http://www.gateway.ingenta.com). IRAS is an affiliate of the American Association for the Advancement of Science (http://www.aaas.org) and holds symposia at its annual meetings.

The Institute for Biblical and Scientific Studies (http://bibleandscience.com)

"The Institute for Biblical and Scientific Studies is a non-profit tax-exempt organization interested in the areas of Bible and science. The goals of the Institute are: To educate people about Bible and Science, and to do research in Bible and Science." The institute publishes an interesting E-mail newsletter semi-weekly.

The John Templeton Foundation (http://www.templeton.org)

The foundation pursues "new insights at the boundary between theology and science through a rigorous, open-minded and empirically focused methodology, drawing together talented representatives from a wide spectrum of fields of expertise. Using 'the humble approach,' the Foundation typically seeks to focus the methods and resources of scientific inquiry on topical areas which have spiritual and theological significance ranging across the disciplines from cosmology to healthcare."

Metanexus Institute on Science and Religion (http://www.metanexus.org)

"[The institute] advances research, education and outreach on the constructive engagement of science and religion. We seek to create an enduring intellectual and social movement by collaborating with persons and communities from diverse religious traditions and scientific disciplines. In a spirit of humility and with a deep concern for intellectual rigor, the Metanexus Institute promotes a balanced and exploratory dialogue between science and religion. While mindful of the complexities of this endeavor, we work to develop integrative approaches that enrich the domains of both science and religion.

"The Metanexus Institute continues to encourage dialogue between religion and science through numerous programs and initiatives [including Metanexus Online, a] subscriber based online magazine that is one of the fastest growing venues for research and publication in the field of science and religion." The institute offers Templeton Research Lectures, the Local Societies Initiative, Spiritual Transformation Scientific Research Program, courses and lectures, and annual conferences.

Center for Theology and the Natural Sciences (http://www.ctns.org)

"[The center is] an international non-profit membership organization dedicated to research, teaching and public service. It focuses on the relation between the natural sciences including physics, cosmology, evolutionary and molecular biology, as well as technology and the environment, and Christian theology and ethics.

"As an Affiliate of the Graduate Theological Union (GTU) in Berkeley, California, CTNS offers courses at the doctoral and seminary levels in order to bring future clergy and faculty to a greater awareness of this important interdisciplinary work." It publishes the journal *Theology and Science*.

References

Aizawa, S. I. 2001. "Bacterial Flagella and Type-III Secretion Systems." *FEMS Microbiology Letters* 202: 157–64.

Altshuler, Edward E., and Derek S. Linden. 1999. "Design of Wire Antennas Using Genetic Algorithms." In *Electromagnetic Optimization by Genetic Algorithms*, edited by Y. Rahmat-Samii and E. Michielssen, 211–48. New York: Wiley.

Andrade, A., et al. 2002. "Expression and Characterization of Flagella in Nonmotile Enteroinvasive *Escherichia coli* Isolated from Diarrhea Cases." *Infection and Immunity* 70: 5882–86.

Anonymous. 2003a. "DNA Structural Analysis of Sequenced Microbial Genomes." http://www.cbs.dtu.dk/services/GenomeAtlas/. Accessed 27 October 2003.

Anonymous. 2003b. "Taking up Serpents." Part 2. http://atheism.about.com/library/weekly/aa081199b.htm. Accessed 2 June 2003.

Anonymous. n.d. "The Wedge Strategy." http://www.antievolution.org/features/wedge.html. Accessed 10 March 2003.

Asai, Y., I. Kawagishi, R. E. Sockett, and M. Homma. 1999. "Hybrid Motor with H(+)- and Na(+)-Driven Components Can Rotate *Vibrio* Polar Flagella by using Sodium Ions." *Journal of Bacteriology* 181: 6332–38.

Asai, Y., T. Yakushi, I. Kawagishi, and M. Homma. 2003. "Ion-Coupling Determinants of Na(+)-Driven and H(+)-Driven Flagellar Motors." *Journal of Molecular Biology* 327: 453–63.

Atkins, P. W. 1994. *The Second Law: Energy, Chaos and Form*. New York: Freeman.

Aunger, Robert. 2002. *The Electric Meme: A New Theory of How We Think*. New York: Free Press.

Babloyantz, A. 1986. *Molecules, Dynamics and Life*. New York: Wiley.

Bakar, Osman. 1987. *Critique of Evolutionary Theory: A Collection of Essays*. Kuala Lumpur: Islamic Academy of Science.

———. 1999. *The History and Philosophy of Islamic Science*. Cambridge, U.K.: Islamic Texts Society.

Barnes, Alfred S. 1939. "The Difference between Natural and Human Flaking on Prehistoric Flint Implements." *American Anthropologist*, new series, 41, 7(1): 99–112.

Barrow, John D., and Frank J. Tipler. 1986. *The Anthropic Cosmological Principle*. Oxford: Oxford University Press.

Begley, Sharon. 1998. "Science Finds God." *Newsweek*, 20 July, pp. 46–52.

Behe, Michael J. 1994. "Molecular Machines: Experimental Support for the Design Inference." Paper presented at the summer meeting of the C. S. Lewis Society, Cambridge University. http://www.arn.org/docs/behe/mb_mm92496.htm. Accessed 15 January 2003.

———. 1996. *Darwin's Black Box: The Biochemical Challenge to Evolution.* New York: Free Press.

———. 1998. "Tracts of Adenosine and Cytidine Residues in the Genomes of Prokaryotes and Eukaryotes." *DNA Sequence* 8: 375–83.

———. 1999. Foreword. In *Intelligent Design: The Bridge between Science and Religion,* by William A. Dembski, 9–12. Downers Grove, Ill.: InterVarsity.

———. 2000. "Self-Organization and Irreducibly Complex Systems: A Reply to Shanks and Joplin." *Philosophy of Science* 67: 155–62.

———. 2001a. "Darwin's Breakdown: Irreducible Complexity and Design at the Foundations of Life." In *Signs of Intelligence: Understanding Intelligent Design,* edited by William A. Dembski and J. M. Kushiner. Grand Rapids, Mich.: Brazos.

———. 2001b. "Reply to My Critics." *Biology and Philosophy* 16: 685–709.

Behe, Michael J., and G. Felsenfeld. 1981. "Effects of Methylation on a Synthetic Polynucleotide: The B-Z Transition in Poly(dG-m5dC).poly(dG-dm5C)." *Proceedings of the National Academy of Science* 78: 1619–23.

Bentley, Peter, ed. 1999. *Evolutionary Design by Computers.* San Francisco: Morgan Kaufmann.

Berg, H. C. 2002. "The Rotary Motor of Bacterial Flagella." *Annual Review of Biochemistry* 72: 19–54.

Berger, James O. 1980. *Statistical Decision Theory: Foundations, Concepts, and Methods.* New York: Springer-Verlag.

Berlekamp, E. R., J. H. Conway, and R. K. Guy. 2003. *Winning Ways, for Your Mathematical Plays.* 2d ed. Natick, Mass.: Peters.

Berlinski, David. 2002. "Has Darwin Met His Match?" *Commentary* 14 (December): 31–41.

———. 2003. "A Scientific Scandal." *Commentary* (April): 29–36.

Berlinksi, David, and his critics. 2003a. "Controversy: Darwinism versus Intelligent Design." *Commentary* 115 (March): 9–31.

———. 2003b. "Controversy: A Scientific Scandal?" *Commentary* 116 (July–August): 26.

Berry, R. M., and J. P. Armitage. 1999. "The Bacterial Flagella Motor." *Advances in Microbial Physiology* 41: 291–337.

Blackmore, Susan. 1999. *The Meme Machine.* Oxford: Oxford University Press.

Boesch, C., and H. Boesch. 1983. "Optimization of Nut-Cracking with Natural Hammers by Wild Chimpanzees." *Behaviour* 83: 265–86.

Boolos, George S., and Richard C. Jeffrey. 1989. *Computability and Logic.* 3d ed. New York: Cambridge University Press.

Boswell, James. 1983. *Life of Johnson.* Oxford: Oxford University Press.

Bowler, Peter. 1989. *Evolution: The History of an Idea.* Berkeley: University of California Press.

Bradley, Walter L., and Charles B. Thaxton. 1994. "Information and the Origin of Life." In *The Creation Hypothesis: Scientific Evidence for the Intelligent Designer,* edited by J. P. Moreland. Downers Grove, Ill.: InterVarsity.

Brooks, Daniel R., and E. O. Wiley. 1988. *Evolution As Entropy: Toward a Unified Theory of Biology.* 2d ed. Chicago: University of Chicago Press.

Brush, A. 2000. "Evolving a Protofeather and Feather Diversity." *American Zoologist* 40: 631–39.

Bryant, H. N., and A. P. Russell. 1992. "The Role of Phylogenetic Analysis in the Inference of Unpreserved Attributes of Extinct Taxa." *Philosophical Transactions of the Royal Society of London B* 337: 405–18.

Bugge, T. H., K. W. Kombrinck, M. J. Flick, C. D. Daugherty, M. J. Danton, and J. L.

Degan. 1996. "Loss of Fibrinogen Rescues Mice for the Pleiotropic Effects of Plasminogen Deficiency." *Cell* 87: 709–19.

Bunge, Mario. 1996. *Finding Philosophy in Social Science*. New Haven, Conn.: Yale University Press.

Burgers, P., and L. M. Chiappe. 1999. "The Wing of Archaeopteryx as a Primary Thrust Generator." *Nature* 399: 60–62.

Burgers, P., and K. Padian. 2001. "Why Thrust and Ground Effect Are More Important Than Lift in the Evolution of Sustained Flight." In *New Perspectives on the Origin and Evolution of Birds: Proceedings of the International Symposium in Honor of John H. Ostrom*, edited by J. Gauthier and L. F. Gall, 351–61. New Haven, Conn.: Peabody Museum of Natural History.

Burton, Thomas. 1993. *Serpent Handling Believers*. Knoxville: University of Tennessee Press.

Byl, J. 1989. "Self-Reproduction in Small Cellular Automata." *Physica D* 34: 295–99.

Calvin, William H. 1996. *The Cerebral Code: Thinking a Thought in the Mosaics of the Mind*. Cambridge, Mass.: MIT Press.

Camazine S., J. L. Deneubourg, N. R. Franks, J. Sneyd, G. Theraulaz, and E. Bonabeau. 2001. *Self-Organization in Biological Systems*. Princeton, N.J.: Princeton University Press.

Carr, Bernard, and Martin Rees. 1979. "The Anthropic Principle and the Structure of the Physical World." *Nature* 278: 605–12.

Carter, Brandon. 1990. "Large Number Coincidences and the Anthropic Principle in Cosmology." In *Confrontation of Cosmological Theory with Astronomical Data*, edited by M. S. Longair, 291–98. Dordrecht, Germany: Reidel; reprinted in *Physical Cosmology and Philosophy*, edited by John Leslie, 61–68. New York: Macmillan.

Cascales, E., R. Lloubes, and J. N. Sturgis. 2001. "The TolQ-TolR Proteins Energize TolA and Share Homologies with the Flagellar Motor Proteins MotA-MotB." *Molecular Microbiology* 42: 795–807.

Cavalier-Smith, Tom. (1997). "The Blind Biochemist." *Tree* 12: 163–64.

Chellapilla, Kumar, and David B. Fogel. 1999. "Co-Evolving Checkers Playing Programs Using Only Win, Lose or Draw." Paper presented at SPIE's AeroSense '99: Applications and Science of Computational Intelligence II, Orlando, Fla., 5–9 April.

Chen, P. J., Z. M. Dong, and S. M. Zhen. 1998. "An Exceptionally Well-Preserved Theropod Dinosaur from the Yixian Formation of China." *Nature* 391: 147–52.

Chiappe, L. M., S. Ji, Q. Ji, and M. A. Norell. 1999. "Anatomy and Systematics of the *Confuciusornithidae* (Theropoda: Aves) from the Late Mesozoic of Northeastern China." *Bulletin of the American Museum of Natural History* 242: 1–89.

Chimpanzee Cultures. 2003. http://138.251.146.69/cultures3/header.html. Accessed 25 October 2003.

Chiprout, Eli. 2003. "A Critique of the Design Inference." http://members.cox.net/chiprout/DesignInference/Demski.htm.

Chure, D. J. 2001. "The Wrist of Allosaurus (Saurischia: Theropoda), with Observations on the Carpus in Theropods." In *New Perspectives on the Origin and Evolution of Birds: Proceedings of the International Symposium in Honor of John H. Ostrom*, edited by J. Gauthier and L. F. Gall, 283–300. New Haven, Conn.: Peabody Museum of Natural History.

Clark, W. R. 1995. *At War Within: The Double-Edged Sword of Immunity*. Oxford: Oxford University Press.

Cohen, Menachem. 2000. "The Religious and the Scientific Aspects of the Debate on the Codes Hidden in the Torah at Equidistant Letter Sequences." www.talkreason.org/articles/cohen.cfm. Accessed 15 October 2002.

Coues, E. 1871. "On the Mechanism of Flexion and Extension in Birds' Wings." *Proceedings of the American Association for the Advancement of Science* 20: 278–84.

Covington, Dennis. 1996. *Salvation on Sand Mountain: Snake Handling and Redemption in Southern Appalachia.* New York: Penguin.

Coyne, J. A. 1996. "God in the Details." *Nature* 383: 227–28.

Cromer, Alan. 1993. *Uncommon Sense: The Heretical Nature of Science.* New York: Oxford University Press.

Cziko, Gary. 2000. *The Things We Do: Using the Lessons of Bénard and Darwin to Understand the What, How, and Why of Our Behavior.* Cambridge, Mass.: MIT Press.

Darwin, Charles. 1859. *The Origin of Species by Means of Natural Selection or, the Preservation of Favored Races in the Struggle for Life.* 6th ed. New York: Burt.

———. 1871. *The Descent of Man and Selection in Relation to Sex.* London: Murray.

Dawkins, Richard. 1986. *The Blind Watchmaker: Why the Evidence of Evolution Reveals a Universe without Design.* New York: Norton.

Dean, Geoffrey. 1986–87. "Does Astrology Need to Be True?" Part 1: "A Look at the Real Thing." *Skeptical Inquirer* 11 (winter): 166–84. Part 2: "The Answer Is No." *Skeptical Inquirer* 11 (spring): 257–73.

Dembski, William A. 1991. "Randomness by Design." http://www.designinference.com/documents/2002.09.rndmnsbydes.pdf. Accessed 3 June 2003.

———. 1994. "On the Very Possibility of Intelligent Design." In *The Creation Hypothesis: Scientific Evidence for the Intelligent Designer,* edited by J. P. Moreland. Downers Grove, Ill.: InterVarsity.

———. 1998a. *The Design Inference: Eliminating Chance through Small Probabilities.* New York: Cambridge University Press.

———. 1998b. "The Explanatory Filter: A Three-Part Filter for Understanding How to Separate and Identify Cause from Intelligent Design." http://www.origins.org/articles/dembski_explanfilter.html.

———. 1998c. "The Intelligent Design Movement." *Cosmic Pursuit.* http://sapiensweb.free.fr/articles/2-dembski.htm.

———. 1998d. Introduction. In *Mere Creation: Science, Faith and Intelligent Design.* Downers Grove, Ill.: InterVarsity.

———, ed. 1998e. *Mere Creation: Science, Faith and Intelligent Design.* Downers Grove, Ill.: InterVarsity.

———. 1998f. "Science and Design." *First Things* 86 (October).

———. 1998g. Untitled document. http://www.firstthings.com/ftissues/ft9810/dembski.html.

———. 1999. *Intelligent Design: The Bridge between Science and Religion.* Downers Grove, Ill.: InterVarsity.

———. 2000 "Conservatives, Darwin and Design: An Exchange" *First Things* 107 (November): 23–31.

———. 2001a. "Is Intelligent Design a Form of Natural Theology?" http://www.designinference.com/documents/2001.03.ID_as_nat_theol.htm.

———. 2001b. "Signs of Intelligence: A Primer on the Discernment of Intelligent Design." In *Signs of Intelligence: Understanding Intelligent Design,* edited by William A. Dembski and J. M. Kushiner, 171–72. Grand Rapids, Mich.: Brazos.

———. 2001c. "What Intelligent Design Is Not." In *Signs of Intelligence: Understanding Intelligent Design,* edited by William A. Dembski and J. M. Kushiner, 7–24. Grand Rapids, Mich.: Brazos.

———. 2002a. "Evolution's Logic of Credulity: An Unfettered Response to Allen Orr." www.designinference.com/documents/2002.12.Unfettered_Resp_to_Orr.htm. Accessed 26 December 2002.

———. 2002b. *No Free Lunch: Why Specified Complexity Cannot Be Purchased without Intelligence.* Lanham, Md.: Rowman and Littlefield.

————. 2002c. "Statement by William A. Dembski on the Scientific Status of Intelligent Design: Edward Kennedy—Expert on Science?" http://www.designinference.com/documents/2002.03.kennedy_ on_ID.htm.

————. 2002d. Untitled document. www.arn.org/cgi-bin/ubb/ultimatebb.cgi?ubb= get_topic;f=13;t=000483. Accessed 26 December 2002.

————. 2003. "Still Spinning Just Fine: A Response to Ken Miller." http://www.designinference.com/documents/2003.02.Miller_Response.htm. Accessed 20 June 2003.

Dembski, William A., and J. M. Kushiner, eds. 2001. *Signs of Intelligence: Understanding Intelligent Design*. Grand Rapids, Mich.: Brazos.

Dennett, Daniel C. 1995. *Darwin's Dangerous Idea: Evolution and the Meanings of Life*. New York: Simon and Schuster.

Dial, K. P. 2003. "Wing-Assisted Incline Running and the Evolution of Flight." *Science* 299: 402–4.

Dial, K. P., S. R. Kaplan, G. E. Goslow, Jr., and F. A. Jenkins. 1988. "A Functional Analysis of the Primary Upstroke and Downstroke Muscles in the Domestic Pigeon (*Columba livia*) during Flight." *Journal of Experimental Biology* 134: 1–16.

Dicke, Robert H. 1961. "Dirac's Cosmology and Mach's Principle." *Nature* 192: 440–41.

Dirac, Paul A. M. 1937. "The Cosmological Constants." *Nature* 139: 323–24.

Ditfurth, Hoimar von. 1982. *The Origins of Life: Evolution As Creation*, translated by Peter Heinegg. San Francisco: Harper and Row.

Dobzhansky, Theodosius. 1964. "Biology, Molecular and Organismic," *American Zoologist* 4: 449.

Doolittle, R. F. 1983. "Probability and the Origin of Life." In *Scientists Confront Creationism*, edited by L. R. Godfrey, 85–97. New York: Norton.

Doroit, R. 1997. "Scientist's Bookshelf—Darwin's Black Box: The Biochemical Challenge to Evolution." *American Scientist* (September–October).

Downing, H., and R. L. Jeanne. 1990. "The Regulation of Complex Building Behaviour in the Paper Wasp, *Polistes fustatus* (Insecta, Hymenoptera, Vespidae)." *Animal Behavior* 39: 105–24.

Drange, Theodore M. 1998. "The Fine-Tuning Argument." www.infidels.org/library/modern/theodore_drange/tuning.html. Accessed 26 December 2002.

Drosnin, M. 1998. *The Bible Code*. New York: Touchstone.

Eaves-Pyles, T. D., H. R. Wong, K. Odoms, and R. B. Pyles. 2001. "Salmonella Flagellin-Dependent Proinflammatory Responses Are Localized to the Conserved Amino and Carboxyl Regions of the Protein." *Journal of Immunology* 167: 7009–16.

Eddington, Arthur S. 1923. *The Mathematical Theory of Relativity*. London: Cambridge.

Edelman, Gerald M. 1992. *Bright Air, Brilliant Fire: On the Matter of the Mind*. New York: Basic Books.

Edis, Taner. 1994. "Islamic Creationism in Turkey," *Creation/Evolution* 34: 1–12.

————. 1998a. "How Gödel's Theorem Supports the Possibility of Machine Intelligence." *Minds and Machines* 8: 251–62.

————. 1998b. "Taking Creationism Seriously." *Skeptic* 6, no. 2: 56–65.

————. 1999. "Cloning Creationism in Turkey," *Reports of the National Center for Science Education* 19: 6, 30–35.

————. 2001. "Darwin in Mind: 'Intelligent Design' Meets Artificial Intelligence." *Skeptical Inquirer* 25, no. 2: 35–39.

————. 2002. *The Ghost in the Universe: God in Light of Modern Science*. Buffalo, N.Y.: Prometheus.

————. 2003. "A World Designed by God: Science and Creationism in Contemporary Islam." In *Science and Religion: Are They Compatible?* edited by Paul Kurtz, 117–25. Buffalo, N.Y.: Prometheus.

Ellis, George. 1993. *Before the Beginning: Cosmology Explained*. London: Boyars/ Bowerdean.

Elsberry, Wesley R. 1999. "Review of *The Design Inference* by William A. Dembski." *Reports of the National Center for Science Education* 19, no. 2: 32–35.

———. 2000. "What Does 'Intelligent Design by Proxy' Do for the Design Inference?" www.talkreason.org/articles/wre_id_proxy.cfm. Accessed 15 October 2002.

Elsberry, Wesley, and Jeffrey Shallit. 2002. "Information Theory, Evolutionary Computation, and Dembski's 'Complex Specified Information.'" www.talkreason.org/ articles/eandsdembski.pdf. Date accessed 2 December 2004.

Evans, H. E. 1966. "The Behaviour Patterns of Solitary Wasps." *Annual Review of Entomology* 11: 123–54.

Everett, Hugh, III. 1957. "'Relative State' Formulation of Quantum Mechanics." *Reviews of Modern Physics* 29: 454–62.

Fauconnier, Gilles, and Mark Turner. 2002. *The Way We Think: Conceptual Blending and the Mind's Hidden Complexities*. New York: Basic Books.

Feduccia, A., and H. B. Tordoff. 1979 "Feathers of *Archaeopteryx* Asymmetric Vanes Indicate Aerodynamic Function." *Science* 203: 1021–22.

Fisher, H. I. 1957. "Bony Mechanism of Automatic Flexion and Extension in the Pigeon's Wing." *Science* 46: 446.

Fitelson, Branden, Christopher Stephens, and Elliott Sober. 1999. "How Not to Detect Design—Critical Notice: William A. Dembski, *The Design Inference*." *Philosophy of Science* 66, no. 3: 472–88; reprinted in *Intelligent Design Creationism and Its Critics: Philosophical, Theological, and Scientific Perspectives*, edited by Robert T. Pennock. New York: MIT Press, 2001.

Fogel, David B. 2000. *Evolutionary Computation: Toward a New Philosophy of Machine Intelligence*. 2d ed. New York: IEEE Press.

Forrest, Barbara. 2001. "The Wedge at Work: How Intelligent Design Creationism Is Wedging Its Way into the Cultural and Academic Mainstream." In *Intelligent Design Creationism and Its Critics: Philosophical, Theological, and Scientific Perspectives*, edited by Robert T. Pennock. New York: MIT Press.

Fortey, R. 1997. *Life—An Unauthorised Biography: A Natural History of the First 4,000,000,000 Years of Life on Earth*. London: Flamingo.

Friedlander, Michael. 1995. *At the Fringes of Science*. Boulder, Colo.: Westview.

Futuyma, Douglas. 1998. *Evolutionary Biology*. 3d ed. Sunderland, Mass.: Sinauer Associates.

Gács, P. 1986. "Randomness and Probability—Complexity of Description." In *Encyclopedia of Statistical Sciences*, 7:551–55. New York: Wiley.

Gadow, H. 1888–93. *Vogel: Aves. Bronn's Klassen und Ordnungen des Their-Reichs*. Leipzig: C. F. Winter'sche Verlagshandlung.

Gajardo, A., A. Moreira, and E. Goles. 2002. "Complexity of Langton's Ant." *Discrete Applied Mathematics* 117: 41–50.

Gardner, Martin. 1986. "WAP, SAP, PAP, and FAP." *New York Review of Books* 23, no. 8: 22–25.

Gauthier, J. A. 1986. "Saurischian Monophyly and the Origin of Birds." In *Memoirs of the California Academy of Sciences*. Volume 8, *The Origin of Birds and the Evolution of Flight*, edited K. Padian, 1–55.

Gauthier, J. A., and K. Padian. 1985. "Phylogenetic, Functional, and Aerodynamic Analyses of the Origin of Birds and Their Flight." In *The Beginnings of Birds: Proceedings of the International Archaeopteryx Conference*, edited by M. K. Hecht, J. H. Ostrom, G. Viohl, and P. Wellnhofer, 185–97. Willibaldsburg, Eichstätt, Germany: Freunde des Jura Museums Eichstätt.

George, J. C., and A. J. Berger. 1966. *Avian Myology*. New York: Academic Press.

Gilmore, C. W. 1915. "On the Forelimb of *Allosaurus fragilis*." *Proceedings of the United States National Museum* 49: 501–13.

Giron, J. A., A. G. Torres, E. Freer, and J. B. Kaper. 2002. "The Flagella of Enteropathogenic *Escherichia coli* Mediate Adherence to Epithelial Cells." *Molecular Microbiology* 44: 361–79.

Gishlick, Alan. D. 2001a. "Evidence for Muscular Control of Avian Style Automatic Extension and Flexion of the Manus in the Forearm of Maniraptors." *Journal of Vertebrate Paleontology* 21 (supplement): 54A–55A.

———. 2001b. "The Function of the Forelimb and Manus of *Deinonychus antirrhopus* and Its Importance for the Origin of Avian Flight." In *New Perspectives on the Origin and Evolution of Birds: Proceedings of the International Symposium in Honor of John H. Ostrom*, edited by J. Gauthier and L. F. Gall, 301–18. New Haven: Peabody Museum of Natural History.

———. 2001c. "Predatory Behaviour in Maniraptors." In *Palaeobiology II*, edited by D. E. G. Briggs and P. R. Crowther, 414–17. Malden, Mass.: Blackwell.

Goles, E., and M. Margenstern. 1996. "Sand Pile As a Universal Computer." *International Journal of Modern Physics C* 7, no. 2: 113–22.

Goslow, G. E., Jr., K. P. Dial, and F. A. Jenkins, Jr. 1989. "The Avian Shoulder: An Experimental Approach." *American Zoologist* 29: 287–301.

———. 1990. "Bird Flight: Insights and Complications." *BioScience* 40: 108–15.

Gould, Stephen Jay. 1993. "An Earful of Jaw." In *Eight Little Piggies*, 95–108. New York: Norton.

Gould, Stephen Jay, and E. S. Vrba. 1982. "Exaptation: A Missing Term in the Science of Form." *Paleobiology* 8: 4–15.

Grassé, P. 1959. La Reconstruction du nid et le coordinations inter-individuelles chez *Bellicositermes natalensis* et *Cubitermes* sp. La Théorie de la stigmergie: essai d'interpretation du comportement des termites constructeurs. *Insectes Sociaux* 6: 41–81.

Gratzer, W. 2000. *The Undergrowth of Science: Delusion, Self-Deception and Human Frailty*. Oxford: Oxford University Press.

Gregory, Simon, et al. 2002. "A Physical Map of the Mouse Genome." *Nature* 418: 743–50

Grinsell, Leslie V. 1976. "Flint Arrowheads, Stone Axes, and Other Small Objects." In *Folklore of Prehistoric Sites in Britain*, 71–72. London: David and Charles.

Guerrero, R., et al. 1986. "Predatory Prokaryotes: Predation and Primary Consumption Evolved in Bacteria." *Proceedings of the National Academy of Sciences* 83: 2138–42.

Guth, Alan. 1981. "Inflationary Universe: A Possible Solution to the Horizon and Flatness Problems." *Physical Review D* 23: 347–56.

Haldane, J. B. S. 1928. "On Being the Right Size." In *Possible Worlds and Other Papers*, 20–28. New York: Harper; http://irl.cs.ucla.edu/papers/right-size.html. Accessed 22 October 2003.

Harris, R., and D. Elder. 2002. "Actin and Flagellin May Have an N-Terminal Relationship." *Journal of Molecular Evolution* 54: 283–84.

Hart, I. B. 1961. *The World of Leonardo da Vinci: Man of Science, Engineer and Dreamer of Flight*. New York: Viking.

Hasofer, A. Michael. 1998. "A Statistical Critique of the Witztum et al. Paper." www.nctimes.net/~mark/fcodes/hasofer.htm. Accessed 15 October 2002.

Haught, John F. 2000. *God after Darwin: A Theology of Evolution*. Boulder, Colo.: Westview.

———. 2003. *Deeper Than Darwin: The Prospect for Religion in an Age of Evolution*. Boulder, Colo.: Westview.

Hedges, Robert. 2003. "Puzzling out the Past." *Nature* 422: 667.

Heeren, F. 2000. "The Deed Is Done." *American Spectator* 33 (December–January): 28.

Hirvensalo, M. 2001. *Quantum Computing*. New York: Springer-Verlag.

Hochrein, Michael J. 2003. *A Bibliography Related to Crime Scene Interpretation with Emphasis in Forensic Geotaphonomic and Forensic Archaeological Field Techniques*. St. Louis, Mo.: Federal Bureau of Investigation

Hood, David W. 2003. "Folk Culture in North East Scotland: An Overview." North East Folk Lore Archive. http://www.nefa.net/archive/hood/hood2.htm. Accessed 3 June 2003.

Hoyle, F. 1954. "In Nuclear Reactions Occurring in Very Hot Stars." Part 1, "The Synthesis of Elements from Carbon to Nickel." *Astrophysical Journal* 1 (supplement): 121–46.

Hoyle, F., D. N. F. Dunbar, W. A. Wensel, and W. Whaling. 1953. "A State of C_{12} Predicted from Astronomical Evidence." *Physical Review Letters* 92: 1095.

Hudson, G. E., and P. J. Lanzillotti. 1955. "Gross Anatomy of the Wing Muscles in the Family *Corvidae*." *American Midland Naturalist* 53: 1–44.

Hueck, C. J. 1998. "Type-III Protein Secretion Systems in Bacterial Pathogens of Animals and Plants." *Microbiology and Molecular Biology Reviews* 62: 379–433.

Ikeda, Michael, and Bill Jefferys. 2001. "The Anthropic Principle Does Not Support Supernaturalism." http://quasar.as.utexas.edu/anthropic.html. Accessed 30 April 2001; www.talkreason.org/articles/super.cfm. Accessed 26 December 2002.

Ingermanson, Randall. 1999. *Who Wrote The Bible Code? A Physicist Probes the Current Controversy*. Colorado Springs.: WaterBrook.

Jenkins, F. A. 1993. "The Evolution of the Avian Shoulder Joint." *American Journal of Science* 293: A253–67.

Ji, Q., P. J. Currie, M. A. Norell, and S. A. Ji. 1998. "Two Feathered Dinosaurs from Northeastern China." *Nature* 393: 753–61.

Johnson, Phillip E. 1993. *Darwin on Trial*, 2d ed. Downers Grove, Ill.: InterVarsity.

———. 1998. "Sinking the Battleship." In *Mere Creation: Science, Faith and Intelligent Design*, edited by William A. Dembski. Downers Grove, Ill.: InterVarsity.

———. 2000. *The Wedge of Truth: Splitting the Foundations of Naturalism*. Downers Grove, IL: InterVarsity.

Junker, R., and Scherer, S. 1988. *Entstehung und Geschichte der Lebewesen*. Giessen, Germany: Weyel Lehrmittelverlag.

Kari, L. 1997. "DNA Computing: Arrival of Biological Mathematics." *Mathematical Intelligencer* 19, no. 2: 9–22.

Karsai, Istvan, and Zsoltan Pénzes. 1993. "Comb Building in Social Wasps: Self-Organization and Stigmergic Script." *Journal of Theoretical Biology* 161: 505–25.

———. 1998. "Nest Shapes in Paper Wasps: Can the Variability of Forms Be Deduced from the Same Construction Algorithm?" *Proceedings of the Royal Society London B* 265: 1261–68.

———. 2000. "Optimality of Cell Arrangement and Rules of Thumb of Cell Initiation in *Polistes dominulus*: A Modeling Approach." *Behavioral Ecology* 11: 387–95.

Karsai, Istvan, and G. Theraulaz. 1995. "Nest Building in a Social Wasp: Postures and Constraints (Hymenoptera: Vespidae)." *Sociobiology* 26: 83–114.

Kauffman, Stuart. 1995. *At Home in the Universe*. New York: Oxford University Press.

———. 2000. *Investigations*. New York: Oxford University Press.

Kazanas, D. 1980. "Dynamics of the Universe and Spontaneous Symmetry Breaking." *Astrophysical Journal* 241: L59–63.

Kimbrough, David. 2002. *Taking Up Serpents: A History of Snake Handling.* Macon, Ga.: Mercer University Press.

Kirchherr, W., M. Li, and P. Vitányi. 1997. "The Miraculous Universal Distribution." *Mathematical Intelligencer* 19, no. 4: 7–15.

Kirov, S. M., et al. 2002. "Lateral Flagella and Swarming Motility in *Aeromonas* Species." *Journal of Bacteriology* 184: 547–55.

Knoll, A. H. 2003. *Life on a Young Planet: The First Three Billion Years of Evolution on Earth.* Princeton, N.J.: Princeton University Press.

Kolmogorov, A. N. 1965. "Three Approaches to the Quantitative Definition of Information." *Problems in Information Transmission* 1: 1–7; *International Journal of Computational Mathematics* 2 (1968): 157–68.

Koons, Robert C. 2001. Unpublished introduction of William A. Dembski. Conference on Design, Self-Organization, and the Integrity of Creation, Calvin College, Grand Rapids, Mich., 25 May.

Korthof, Gert. 2001. "Does Life Look Like or Unlike Evolution?" http://home.wxs.nl/~gkorthof/kortho41.htm. Accessed December 2002.

———. 2002. "A Chemist's View of Life: Ultimate Reductionism and Dissent." http://home.wxs.nl/~gkorthof/korthof56.htm. Accessed January 2003.

———. 2003. "Independent Origin and the Facts of Life." http://home.wxs.nl/~gkorthof/korthof59.htm. Accessed June 2003.

Kuhnert, L., K. I. Agladze, and V. I. Krinsky. 1989. "Image Processing Using Light-Sensitive Chemical Waves." *Nature* 337: 244–47.

Laplace, P. S. 1951. *A Philosophical Essay on Probabilities*, translated by F. W. Truscott and F. L. Emory. New York: Dover.

Larson, Edward J. 1989. *Trial and Error: The American Controversy over Creation and Evolution.* New York: Oxford University Press.

Larson, Edward J., and Larry Witham. 1998. "Leading Scientists Still Reject God." *Nature* 394: 313.

Laudan, Larry. 1988. "The Demise of the Demarcation Problem." In *But Is It Science? The Philosophical Question in the Creation/Evolution Controversy*, edited by Michael Ruse, 337–50. Buffalo, N.Y.: Prometheus.

Lawren, Bill. 1992. "The Case of the Ghost Molecules." *Omni* 14 (June): 50–52, 73–74.

Levi, Barbara Goss. 1973. "Anomalous Water: An End to the Anomaly," *Physics Today* 26 (October): 19–20.

Levin, L. A. 1984. "Randomness Conservation Inequalities: Information and Independence in Mathematical Theories." *Information and Control* 61: 15–37.

Lewis, David. 1986. *On the Plurality of Worlds.* Oxford: Blackwell.

Lewontin, R. 2001. *The Triple Helix: Gene, Organism, and Environment.* Cambridge: Harvard University Press.

Li, M., and P. M. B. Vitányi. 1988. "Two Decades of Applied Kolmogorov Complexity." In *Proceedings of the 3rd Structures in Complexity Theory Conference,* 80–101. New York: IEEE Press.

Linde, Andrei. 1982. "A New Inflationary Universe Scenario: A Possible Solution of the Horizon, Flatness, Homogeneity, Isotropy, and Primordial Monopole Problems." *Physics Letters* 108B: 389–92.

———. 1990. *Particle Physics and Inflationary Cosmology.* New York: Academic Press.

———. 1994. "The Self-Reproducing Inflationary Universe." *Scientific American* 271 (November): 48–55.

Livingstone, David N. 1987. *Darwin's Forgotten Defenders: The Encounter between Evangelical Theology and Evolutionary Thought.* Grand Rapids, Mich.: Eerdman.

Livio, M., D. A. Hollowel, A. Weiss, and J. Truran. 1989. "The Anthropic Significance of an Excited State of ^{12}C." *Nature* 349: 281–84.

Lucas, J. R. 1961. "Minds, Machines, and Gödel." *Philosophy* 36: 112.

Macnab, R. M. 1992. "Genetics and Biogenesis of Bacterial Flagella." *Annual Review of Genetics* 26: 131–58.

———. 1999. "The Bacterial Flagellum: Reversible Rotary Propeller and Type-III Export Apparatus." *Journal of Bacteriology* 181: 7149–53.

Madsen, J. H. 1976. "*Allosaurus fragilis*: A Revised Osteology." *Utah Geological Survey Bulletin* 109: 1–163.

Maeterlinck, M. 1927. *The Life of the White Ant*. London: Allen and Unwin.

Max Planck Research. 2002. Press release. http://www.mpg.de/bilderBerichte Dokumente/dokumentation/pressemitteilungen/2002/pri0243.pdf.

Mayr, Ernst. 1982. *The Growth of Biological Thought*. Cambridge.: Harvard University Press.

———. 2001. *What Evolution Is*. New York: Basic Books.

———. 2002. *What Evolution Is*. London: Phoenix.

McClain, J., D. R. Rollo, B. G. Rushing, and C. E. Bauer. 2002. "*Rhodospirillum centenum* Utilizes Separate Motor and Switch Components to Control Lateral and Polar Flagellum Rotation." *Journal of Bacteriology* 184: 2429–38.

McDonald, John D. 2000. "A Reducibly Complex Mousetrap." http://udel.edu/ ~mcdonald/mousetrap.html. Accessed 6 October 2000.

McKay, Brendan. 1997. "Assassinations Foretold in Moby Dick!" http://cs.anu.edu.au/ ~bdm/dilugim/moby.html. Accessed 15 January 2003.

McKay, Brendan, Dror Ben-Natan, Maya Bar-Hillel, and Gil Kalai. 1999. "Solving the Bible Code Puzzle." *Statistical Science* 14, no. 2: 150–73.

McKitrick, M. C. 1991. "Forelimb Myology of Loons (Gaviiformes), with Comments on the Relationship of Loons and Tubenoses (Procellariiformes)." *Zoological Journal of the Linnean Society* 102: 115–52.

Meinersmann, R. J., and K. L. Hiett. 2000. "Concerted Evolution of Duplicate fla Genes in Campylobacter." *Microbiology* 146: 2283–90.

Mercader, J., M. Panger, and C. Boesch. 2002. "Excavation of a Chimpanzee Stone Tool Site in the African Rainforest." *Science* 296: 1452–55.

Miller, Kenneth R. 1999. *Finding Darwin's God: A Scientist's Search for Common Ground between God and Evolution*. New York: Cliff Street Books.

———. 2003. "The Flagellum Unspun: The Collapse of 'Irreducible Complexity.'" In *Debating Design: From Darwin to DNA*, edited by Michael Ruse and William A. Dembski. New York: Cambridge University Press.

Monod, J. 1971. *Chance and Necessity: An Essay on the Natural Philosophy of Modern Biology*. New York: Knopf.

Moreland, J. P., ed. 1994. *The Creation Hypothesis: Scientific Evidence for the Intelligent Designer*. Downers Grove, Ill: InterVarsity.

———. 1999. "Postmodernism and the Intelligent Design Movement." *Philosophia Christi*, 2d series, 1, no. 2: 97–101.

Namba, K., et al. 2003. "Protonic Nanomachine Project." Exploratory Research for Advanced Technology. http://www.npn.jst.go.jp/. Accessed 30 June 2003.

Nasr, Seyyed Hossein. 1989. *Knowledge and the Sacred*. Albany: State University of New York Press.

Nelson, Paul. 2001a. "The Role of Theology in Current Evolutionary Reasoning." In *Intelligent Design Creationism and Its Critics: Philosophical, Theological, and Scientific Perspectives*, edited by Robert T. Pennock. New York: MIT Press.

———. 2001b. "Unfit for Survival. The Fatal Flaws of Natural Selection." In *Signs of Intelligence: Understanding Intelligent Design*, edited by William A. Dembski and J. M. Kushiner. Grand Rapids, Mich.: Brazos.

Nilsson, Dan-E., and Susanne Pelger. 1994. "A Pessimistic Estimate of the Time Required for an Eye to Evolve." *Proceedings of the Royal Society of London* 256: 53–58.

Norberg, U. M. 1990. *Vertebrate Flight: Mechanics Physiology, Morphology, Ecology, and Evolution.* New York: Springer-Verlag.

Nord, Warren A. 1995. *Religion and American Education: Rethinking a National Dilemma.* Chapel Hill: University of North Carolina Press.

———. 1999. "Religion-Free Texts: Getting an Illiberal Education." *Christian Century* 116 (July): 14–21, 711–15.

Numbers, Ronald L. 1992. *The Creationists: The Evolution of Scientific Creationism.* New York: Knopf.

Nunn, D. 1999. "Bacterial Type II Protein Export and Pilus Biogenisis: More Than Just Homologies?" *Trends in Cell Biology* 9: 402–408.

Olsen, Ted. 1998. "They Shall Take Up Serpents." *Christianity Today.* http://www.christianitytoday.com/ch/58h/58h025.html. Accessed 27 May 2003.

Oosawa, F., and S. Hayashi. 1986. "The Loose Coupling Mechanism in Molecular Machines of Living Cells." *Advances in Biophysics* 22: 151–83.

Orr, H. A. 1996–1997. "Darwin v. Intelligent Design (Again)." Review of *Darwin's Black Box: The Biochemical Challenge to Evolution,* by Michael Behe. *Boston Review* 21, no. 6 (December–January): 28–31.

———. 2002. "The Return of Intelligent Design." Review of *No Free Lunch: Why Specified Complexity Cannot Be Purchased without Intelligence,* by William A. Dembski. *Boston Review* 27, nos. 3–4 (Summer): 53–56.

Ostrom, J. H. 1969. "Osteology of *Deinonychus antirrhopus,* an Unusual Theropod from the Lower Cretaceous of Montana." *Bulletin of the Peabody Museum of Natural History* 30: 1–165.

———. 1974. "*Archaeopteryx* and the Origin of Flight." *Quarterly Review of Biology* 49: 27–47.

———. 1976. "*Archaeopteryx* and the Origin of Birds." *Biological Journal of the Linnean Society* 8: 91–182.

———. 1979. "Bird Flight: How Did It Begin?" *American Scientist* 67: 46–56.

———. 1997. "How Bird Flight Might Have Come About." In *Dinofest International,* edited by D. L. Wolberg, E. Stump, and G. D. Rosenberg, 301–10. Washington, D.C.: Academy of Natural Sciences.

Padian, K. 1982. "Macroevolution and the Origin of Major Adaptations: Vertebrate Flight As a Paradigm for the Analysis of Patterns." *Proceedings, Third North American Paleontological Convention:* 387–92.

———. 2001. "Cross-Testing Adaptive Hypotheses: Phylogenetic Analysis and the Origin of Bird Flight." *American Zoologist:* 598–607.

Padian, K., and L. M. Chiappe. 1998a. "The Origin of Birds and Their Flight." *Scientific American* (February): 38–47.

———. 1998b. "The Origin and Early Evolution of Birds." *Biological Reviews of the Cambridge Philosophical Society* 73: 1–42.

Paley, W. 1802. *Natural Theology.* London: Faulder.

Pate, J. L., and L. Y. E. Chang. 1979. "Evidence that Gliding Motility in Prokaryotic Cells Is Driven by Rotary Assemblies in the Cell Envelope." *Current Microbiology* 2: 59–64.

Peacocke, Arthur R. 1986. *God and the New Biology.* London: Dent.

Pennisi, Elizabeth. 2003. "Modernizing the Tree of Life." *Science* 300, no. 5626: 1692–97.

Pennock, Robert T. 1996. "Naturalism, Evidence, and Creationism: The Case of Phillip Johnson." *Biology and Philosophy* 11, no. 4: 543–49.

———. 1999. *Tower of Babel: The Evidence against the New Creationism*. Cambridge, Mass.: MIT Press.

———, ed. 2001. *Intelligent Design Creationism and Its Critics: Philosophical, Theological, and Scientific Perspectives*. New York: MIT Press.

Penrose, Roger. 1989. *The Emperor's New Mind*. Oxford: Oxford University Press.

———. 1994. *Shadows of the Mind*. Oxford: Oxford University Press.

Perakh, Mark. 1998–2000. "B-codes." www.nctimes.net/~mark/fcodes/. Accessed 15 October 2002.

———. 2000. "Vzlyot I Padeniye Bibleyskikh Kodov [The Rise and Fall of the Bible Code]." *Kontinent* 103 (January–March): 240–70.

———. 2001a. "Anthropic Principles—Reasonable and Unreasonable." www.talkreason.org/articles/anthropic.cfm. Accessed 26 December 2002.

———. 2001b. "A Consistent Inconsistency. How Dr. Dembski Infers Design." www.talkreason.org/articles/dembski.cfm. Accessed 26 December 2002.

———. 2002a. "A Free Lunch in a Mousetrap." www.talkreason.org/articles/dem_nfl.cfm. Accessed 26 December 2002.

———. 2002b. "Incompatible Magisteria." www.talkreason.org/articles/magisteria.cfm. Accessed 2 June 2003.

———. 2002c. "A Presentation without Arguments." *Skeptical Inquirer* 26, no. 6: 31.

———. 2002d. "Science in the Eyes of a Scientist." www.talkreason.org/articles/good_bad_science.cfm. Accessed 26 December 2002.

———. 2004a. *Unintelligent Design*. Buffalo, N.Y.: Prometheus.

———. 2004b. "The Rise and Fall of the Bible Code." www.talkreason.org/articles/Codpaper1.cfm. Accessed 12 December 2004. Translation of "Vzlyot I Padeniye Bibleyskikh Kodov," originally published in *Kontinent* 103 (January–March 2000): 240–270.

Perakh, Mark, and Brendan McKay. 1999. "Letter Serial Correlation in Various Texts." http://members.cox.net/marperak/Texts/. Date accessed 2 December 2004.

Perlmutter, S., et al. 1999. "Measurements of Omega and Lambda from 42 High-Redshift Supernovae." *Astrophysical Journal* 517: 565–86.

Plantinga, Alvin. 1991. "When Faith and Reason Clash: Evolution and the Bible." *Christian Scholar's Review* 21, no. 1: 8–32.

Polkinghorne, John. 1998. *Belief in God in an Age of Science*. New Haven, Conn.: Yale University Press.

Poore, S. O., A. Ashcroft, A. Sánchez-Haiman, and G. E. Goslow, Jr. 1997. "The Contractile Properties of the M. Supracoracoideus in the Pigeon and Starling: A Case for Long-Axis Rotation of the Humerus." *Journal of Experimental Biology* 200: 2987–3002.

Poore, S. O., A. Sánchez-Haiman, and G. E. Goslow, Jr. 1997. "Wing Upstroke and the Evolution of Flapping Flight." *Nature* 387: 799–802.

Press, W. H., and Alan P. Lightman. 1983. "Dependence of Macrophysical Phenomena on the Values of the Fundamental Constants." *Philosophical Transactions of the Royal Society of London* A 310: 323–36.

Prum, R. O., and A. H. Brush. 2002. "The Evolutionary Origin and Diversification of Feathers." *Quarterly Review of Biology* 77: 261–95.

———. 2003. "Which Came First, the Feather or the Bird: A Long-Cherished View of How and Why Feathers Evolved Has Now Been Overturned." *Scientific American* 285 (March): 84–93.

Ptashne, M., and A. Gann. 2002. *Genes and Signals*. New York: Cold Spring Harbor.

Raël, Claude Vorilhon. 1986 [1974]. "The Book Which Tells the Truth." In *The Message Given to Me by Extraterrestrials*. Tokyo: AOM Corporation.

Rambidi, N. G., and D. Yakovenchuk. 2001. "Chemical Reaction-Diffusion Implementation of Finding the Shortest Paths in a Labyrinth." *Physical Review E* 63: 26607.

Ratzsch, Del. 2001. *Nature, Design and Science*. Albany: State University of New York Press.

Rau, Phillip. 1929. "Orphan Nests of *Polistes* (Hym.: Vespidae)." *Entomological News* 40: 226–59.

Rayner, J. M. V. 1988. "Form and Function in Avian Flight." *Current Ornithology* 5: 1–66.

Reiss, A., et al. 1998. "Observational Evidence from Supernovae for an Accelerating Universe and a Cosmological Constant." *Astronomical Journal* 116: 1009–38.

Remine, Walter. 1993. *The Biotic Message: Evolution versus Message Theory*. Saint Paul, Minn.: St. Paul Science.

Richter, Reed. 2002. "What Science Can and Cannot Say: The Problems with Methodological Naturalism." *Reports of the National Center for Science Education* 22: 1–2, 18–22.

Ridley, Mark. 1985. *The Problems of Evolution*. Oxford: Oxford University Press.

———. 1996. *Evolution*. 2d ed. Cambridge, Mass.: Blackwell Science.

Rietschel, S. 1985. "Feathers and Wings of *Archaeopteryx* and the Question of Her Flight Ability." In *The Beginnings of Birds: Proceedings of the International Archaeopteryx Conference*, edited by M. K. Hecht, J. H. Ostrom, G. Viohl, and P. Wellnhofer, 251–60. Willibaldsburg, Eichstätt, Germany: Freunde des Jura Museums Eichstätt.

Rolston, Holmes, III. 1999. *Genes, Genesis, and God: Values and their Origins in Natural and Human History*. New York: Cambridge University Press.

Rosenhouse, Jason. 2002. "Probability, Optimization Theory, and Evolution." *Evolution* 56, no. 8: 1721.

Ross, Hugh. 1995. *The Creator and the Cosmos: How the Greatest Scientific Discoveries of the Century Reveal God*. Colorado Springs: Navpress.

———. 1998. *The Genesis Question: Scientific Advances and the Accuracy of Genesis*. Colorado Springs: Navpress.

Rothemund, P. W. K., and E. Winfree. 2000. "The Program-Size Complexity of Self-Assembled Squares." In *Proceedings of the Thirty-second Annual ACM Symposium on the Theory of Computation (STOC)*, 459–468. New York.

Ruse, Michael. 2001. *Can a Darwinian be a Christian? The Relationship between Science and Religion*. New York: Oxford University Press.

———. 2003. *Darwin and Design—Does Evolution Have a Purpose?* Cambridge: Harvard University Press.

Sainsbury, R. M. 1995. *Paradoxes*. 2d ed. New York: Cambridge University Press.

Samatey, F. A., et al. 2000. "Crystal Structure of F41 Fragment of Flagellin." PubMed. http://www.ncbi.nlm.nih.gov/Structure/mmdb/mmdbsrv.cgi?form=6&db=t&Dopt=s&uid=15815.

———. 2001. "Structure of the Bacterial Flagellar Protofilament and Implications for a Switch for Supercoiling." *Nature* 410: 331–37..

Samuel, A. D., J. D. Petersen, and T. S. Reese. 2001. "Envelope Structure of *Synechococcus* sp. WH8113, a Nonflagellated Swimming Cyanobacterium." *BMC Microbiology* 1: 4.

Sandeen, Ernest R. 1970. *The Roots of Fundamentalism: British and American Millenarianism, 1800–1930*. Chicago: University of Chicago Press.

Sanger Institute. Graphical representation of cases of mouse-human synteny. http://www.sanger.ac.uk/Projects/M_musculus/publications/fpcmap–2002/syndata/hm.14.12.html.

Sanz, J. L., L. M. Chiappe, B. P. Perez-Moreno, A. D. Buscalioni, J. J. Moratella, F. Ortega, and F. J. Pouato-Ariza. 1996. "An Early Cretaceous Bird from Spain and Its Implications for the Evolution of Avian Flight." *Nature* 382: 442–45.

Sarfati, Jonathan. 2000. *Refuting Evolution*. Green Forest, Ariz.: Master Books.

Sayin, Ümit, and Aykut Kence. 1999. "Islamic Scientific Creationism." *Reports of the National Center for Science Education* 19, no. 6: 18–29.

Scherer, Siegfried. 1998. "Basic Types of Life." In *Mere Creation: Science, Faith and Intelligent Design*, edited by William A. Dembski. Downers Grove, Ill.: InterVarsity.

Schneider, Thomas D. 2000. "Evolution of Biological Information." *Nucleic Acids Research* 28, no. 14: 2794–99.

———. 2001a. "Effect of Ties on the Evolution of Information by the Ev Program." www.lecb.bcifcrf.gov/~toms/paper/ev/dembski/claimtest.html. Accessed 30 December 2002.

———. 2001b. "Rebuttal to William A. Dembski Posting and to His Book *No Free Lunch*." www.lecb.ncifcrf.gov/~toms/paper/ev/dembski/rebuttal.html. Accessed 30 December 2002.

———. 2002. "Dissecting Dembski's 'Complex Specified Information.'" http://www.lecb.ncifcrf.gov/~toms/paper/ev/dembski/specified.complexity.html. Accessed 24 October 2002.

Schopf, J. W. 1999. *Cradle of Life: The Discovery of Earth's Earliest Fossils.* Princeton, N.J.: Princeton University Press.

Seeley, T. D. 2002. "When Is Self-Organization Used in Biological Systems?" *Biological Bulletin* 202: 314–18.

Sekiya, K., et al. 2001. "Supermolecular Structure of the Enteropathogenic *Escherichia coli* Type-III Secretion System and Its Direct Interaction with the EspA-Sheathlike Structure." *Proceedings of the National Academy of Sciences* 98: 11638–43.

Shah, D. S., and R. E. Sockett. 1995. "Analysis of the motA Flagellar Motor Gene from *Rhodobacter sphaeroides*, a Bacterium with a Unidirectional, Stop-Start Flagellum." *Molecular Microbiology* 17: 961–69.

Shallit, Jeffrey. 2002a. "Anatomy of a Creationist Tall Tale." http://www.math.uwaterloo.ca/~shallit/cw.html.

———. 2002b. "Review of William Dembski, *No Free Lunch: Why Specified Complexity Cannot Be Purchased without Intelligence*." *Biosystems* 66: 93–99.

Shanks, Niall. 2001. "Modeling Biological Systems: The Belousov-Zhabotinski Reaction." *Foundations of Chemistry* 3: 33–53.

Shanks, Niall, and Karl H. Joplin. 1999. "Redundant Complexity: A Critical Analysis of Intelligent Design in Biochemistry." *Philosophy of Science* 66: 268–82.

———. 2000. "Of Mousetraps and Men: Behe on Biochemistry." *Reports of the National Center for Science Education* 20: 25–30.

———. 2001. "Behe, Biochemistry and the Invisible Hand." *Philo* 4: 54–67.

Shufeldt, R. W. 1898. *The Myology of the Raven.* New York: Macmillan.

Simon, Barry. 1998. "The Case against the Code." *Jewish Action* 58, no. 3: 16–24; http://www.wopr.com/biblecodes/The Case.htm. Accessed 25 October 2003.

Smith, Quentin. 1990. "A Natural Explanation of the Existence and Laws of Our Universe." *Australasian Journal of Philosophy* 68: 22–43.

Smolin, Lee. 1992. "Did the Universe Evolve?" *Classical and Quantum Gravity* 9: 173–91.

———. 1997. *The Life of the Cosmos.* Oxford: Oxford University Press.

Sober, Elliott. In press. "The Design Argument." In *Guide to the Philosophy of Religion*, edited by W. Mann. Malden, Mass.: Blackwell.

Spormann, A. M. 1999. "Gliding Motility in Bacteria: Insights from Studies of *Myxococcus xanthus*." *Microbiology and Molecular Biology Reviews* 63: 621–41.

Stark, Rodney, and Roger Finke. 2000. *Acts of Faith: Explaining the Human Side of Religion.* Berkeley: University of California Press.

Steinbock, O., A. Tóth, and K. Showalter. 1995. "Navigating Complex Labyrinths: Optimal Paths from Chemical Waves." *Science* 267: 868–71.

Stenger, Victor J. 1995. *The Unconscious Quantum: Metaphysics in Modern Physics and Cosmology.* Buffalo, N.Y.: Prometheus.

———. 2000. "Natural Explanations for the Anthropic Coincidences." *Philo* 3: 50–67.

————. 2001a. "Anthropic Design: Does the Cosmos Show the Evidence of Purpose?" www.talkreason.org/articles/coincidence.cfm. Accessed 26 December 2002.

————. 2001b. "Natural Explanation for the Anthropic Coincidences." www.talkreason.org/articles/anthro-philo.pdf. Accessed 26 December 2002.

Stevens, Clare. 1998. "A Rebuttal of Behe." http://www.btinternet.com/~clare.stevens/behenot.htm. Accessed 25 August 2001.

Stoeber, Michael, and Hugo Meynell, eds. 1996. *Critical Reflections on the Paranormal.* Albany: State University of New York Press.

Strickberger, Monroe. 2000. *Evolution.* 3d ed. Sudbury, Mass.: Jones and Bartlett.

Sumida, S. S., and C. A. Brochu. 2000. "Phylogenetic Context for the Origin of Feathers." *American Zoologist* 40: 486–503.

Swinburne, Richard. 1998. "Argument from the Fine-Tuning of the Universe." In *Modern Cosmology and Philosophy,* edited by John Leslie, 160–79. Amherst, N.Y.: Prometheus.

Tegmark, Max. 2003. "Parallel Universes." *Scientific American* 288, no. 5: 40–51.

Tellgren, Erik. 2002. "On Dembski's Law of Conservation of Information." www.talkreason.org/articles/dembski_LCI.pdf. Accessed 15 November 2002.

Thagard, Paul. 1992. *Conceptual Revolutions.* Princeton, N.J.: Princeton University Press.

Thaxton, Charles B., Walter L. Bradley, and Roger L. Olson. 1984. *The Mystery of Life's Origin.* Dallas, Tex.: Lewis and Stanley; reprint, 1992.

Thomas, N. A., S. L. Bardy, and K. F. Jarrell. 2001. "The Archaeal Flagellum: A Different Kind of Prokaryotic Motility Structure." *FEMS Microbiology Reviews* 25: 147–74.

Thornhill, R., and David W. Ussery. 2000. "A Classification of Possible Routes in Darwinian Evolution." *Journal of Theoretical Biology* 202: 111–16.

Thorpe, W. H. 1963. *Learning and Instinct in Animals.* London: Methuen.

Tobalske, B. W. 2000. "Biomechanics and Physiology of Gait Selection in Flying Birds." *Physiological and Biochemical Zoology* 73: 736–50.

Tudge, Colin. 2000. *The Variety of Life.* Oxford: Oxford University Press.

Tyson, J. T. 1994. "What Everyone Should Know about the Belousov-Zhabotinski Reaction." In *Frontiers in Mathematical Biology,* edited by S. A. Levin. New York: Springer-Verlag.

Ussery, David W. 1999. "A Biochemist's Response to 'The Biochemical Challenge to Evolution.'" *Bios* 70: 40–45.

Ussery, David W., D. M. Soumpasis, S. Brunak, H. H. Stærfeldt, P. Worning, and A. Krogh. 2002. "Bias of Purine Stretches in Sequenced Genomes." *Computers in Chemistry* 26: 531–41.

Van Till, Howard. 2002. "*E coli* at the No Free Lunchroom. Bacterial Flagella and Dembski's Case for Intelligent Design." www.counterbalance.net/id-hvt/index-body.html. Accessed 5 July 2003.

Vazquez, R. J. 1992. "Functional Osteology of the Avian Wrist and the Evolution of Flapping Flight." *Journal of Morphology* 211: 259–62.

————. 1993. "Functional Morphology of the Avian Wing and Its Implications for *Archaeopteryx* and the Evolution of Birds." Ph.D. dissertation. Yale University.

————. 1994. "The Automating Skeletal and Muscular Mechanism of the Avian Wing (Aves)." *Zoomorphology* 114: 59–71.

Watnick, P. I., et al. 2001. "The Absence of a Flagellum Leads to Altered Colony Morphology, Biofilm Development and Virulence in *Vibrio cholerae* O139." *Molecular Microbiology* 39: 223–35.

Wein, Richard. 2002a. "Not a Free Lunch but a Box of Chocolates." www.talkreason.org/articles/chock_nfl.cfm. Accessed 26 December 2002.

————. 2002b. "Response? What Response? How Dembski Avoided Addressing My Arguments." www.talkreason.org/articles/response.cfm. Accessed 26 December 2002.

Weinberg, Steven. 1989. "The Cosmological Constant Problem." *Reviews of Modern Physics* 61: 1–23.

Wells, Jonathan. 2000. *Icons of Evolution: Science or Myth. Why Much We Teach about Evolution Is Wrong.* Washington, D.C.: Regnery.

West Eberhard, M. J. 1969. "The Social Biology of Polistine Wasps." *Miscellaneous Publications, Museum of Zoology,* 140:1–101. Ann Arbor: University of Michigan.

Weyl, Hermann. 1919. "A New Extension of the Theory of Relativity." *Annalen der Physik* 59: 101.

Wilkins, J. S., and Wesley R. Elsberry. 2001. "The Advantages of Theft over Toil: The Design Inference and Arguing from Ignorance." *Biology and Philosophy* 16: 711–24.

Witham, Larry A. 2002. *Where Darwin Meets the Bible: Creationists and Evolutionists in America.* New York: Oxford University Press.

Witmer, L. M. 1995. "The Extant Phylogenetic Bracket and the Importance of Reconstructing Soft Tissues in Fossils." In *Functional Morphology in Vertebrate Paleontology,* edited by J. J. Thomason, 19–33. New York: Cambridge University Press.

Witztum, Doron, Eliyahu Rips, and Yoav Rosenberg. 1994. "Equidistant Letter Sequences in the Book of Genesis." *Statistical Science* 9, no. 3: 429–38.

Wolpert, David H. 1996a. "The Existence of A *Priori* Distinctions between Learning Algorithms." *Neural Computing* 8, no. 7: 1391–1420.

———. 1996b. "The Lack of A *Priori* Distinctions between Learning Algorithms." *Neural Computing* 8, no. 7: 1341–1390.

———. 2002. "William Dembski's Treatment of the No Free Lunch Theorems Is Written in Jello." www.talkreason.org/articles/jello.cfm. Accessed 26 December 2002.

Wolpert, David H., and William G. Macready. 1997. "The No Free Lunch Theorems for Optimization." *IEEE Transactions on Evolutionary Computation* 1, no. 1: 67–82.

Yahya, Harun. 1997. *Evrim Aldatmacası: Evrim Teorisinin Bilimsel Çöküşü ve Teorinin Ideolojik Arka Planı.* Istanbul: Vural Yayıncılık.

———. 2002. *Fascism: The Bloody Ideology of Darwinism.* Istanbul: Kultur.

Yonekura, K., S. Maki-Yonekura, and K. Namba. 2002. "Growth Mechanism of the Bacterial Flagellar Filament." *Research in Microbiology* 153: 191–97.

Youderian, P., N. Burke, D. J. White, and P. L. Hartzell. 2003. "Identification of Genes Required for Adventurous Gliding Motility in *Myxococcus xanthus* with the Transposable Element Mariner." *Molecular Microbiology* 49: 555–70.

Young, G. M., D. H. Schmiel, and V. L. Miller. 1999. "A New Pathway for the Secretion of Virulence Factors by Bacteria: The Flagellar Export Apparatus Functions As a Protein-Secretion System." *Proceedings of the National Academy of Sciences* 96: 6456–61.

Young, Matt. 1998. "The Bible As a Science Text." *Rocky Mountain Skeptic* 16 (November–December): 2–4.

———. 2000. "The Genesis Question: Scientific Advances and the Accuracy of Genesis." *Rocky Mountain Skeptic* 17 (July): 2–4; www.mines.edu/~mmyoung/BkRevs.htm. Posted 20 November 2001.

———. 2001a. "Intelligent Design Is Neither." Paper presented at the conference "Science and Religion: Are They Compatible?" Atlanta, 9–11 November 2001. www.mines.edu/~mmyoung/DesnConf.pdf, revision 1. Posted 21 February 2002.

———. 2001b. *No Sense of Obligation: Science and Religion in an Impersonal Universe.* Bloomington, Ind.: 1stBooks Library.

———. 2002. "How to Evolve Specified Complexity by Natural Means." Revision 1. www.pcts.org/journal/young2002a.html. Accessed 10 March 2002.

About the Contributors

Co-editor TANER EDIS is assistant professor of physics at Truman State University and a regular researcher at Lawrence Livermore National Laboratory. He is the author of *The Ghost in the Universe: God in Light of Modern Science* (Amherst, N.Y.: Prometheus) and "Darwin in Mind: 'Intelligent Design' Meets Artificial Intelligence" (*Skeptical Inquirer* 25: 35–39), a critique of William Dembski's work. *The Ghost in the Universe* won the Council for Secular Humanism's Morris D. Forkosch Award for the best humanist book of 2002.

Professor Edis is associate editor for physics and astronomy of the *Reports of the National Center for Science Education*, a leading source for critiques of and news about creationism. He earned his Ph.D. from Johns Hopkins University and has published technical papers on physics, philosophy and artificial intelligence, and creationism as well as semi-popular articles on science and religion. He is interested in varieties of creationism (especially Islamic) and presented an invited paper at the first international conference on "Science and Religion: Are They Compatible?" in November 2001. He teaches an interdisciplinary course called "Weird Science," which examines creationism and intelligent design at the level of college juniors and seniors, and gives frequent talks about the scientific failures of intelligent-design creationism. Professor Edis's web site is *http://www2.truman.edu/~edis*.

Co-editor MATT YOUNG formerly held the position of physicist with the National Institute of Standards and Technology (NIST), where he was also chairman of the editorial review board for more than 10 years. He now teaches physics and engineering at the Colorado School of Mines. He remains a NIST Associate and teaches optics and error analysis in an annual short course on laser power and energy measurements. Professor Young is a Fellow of the Optical Society of America and winner of the U.S. Department of Commerce's gold and silver medals for his work in optical communications. He is the author of three books: *No Sense of Obligation: Science and Religion in an Impersonal*

Universe (Bloomington, Ind.: 1stBooks Library); *Optics and Lasers, Including Fibers and Optical Waveguides,* 5th ed. (New York: Springer-Verlag); and *The Technical Writer's Handbook* (Mill Valley, Calif.: University Science Books). All three books are still in print.

Professor Young has published roughly 100 papers, including several in science and religion, in journals ranging from *Applied Optics* to *Skeptical Inquirer* and *Free Inquiry*. In 2001, he presented an invited paper at the first international conference "Science and Religion: Are They Compatible?" and led a workshop at the 2002 conference of the Institute on Religion in an Age of Science.

Formerly, he was a faculty member at the University of Waterloo (Canada) and Rensselaer Polytechnic Institute; a consultant to General Electric, the New York State Energy Commission, and elsewhere; and a visiting scientist at the Weizmann Institute of Science in Rehovot, Israel. He has also served on the board of his local synagogue and as a trustee of the Hillel Council of Colorado, a statewide campus organization for Jewish students. Professor Young's web sites are *http://www.mines.edu/~mmyoung* and *http://www.1stBooks.com/bookview/5559*.

WESLEY R. ELSBERRY is an interdisciplinary researcher in zoology and computer science with a Ph.D. in wildlife and fisheries sciences from Texas A&M University. He is a co-author of peer-reviewed papers in the *Journal of Experimental Biology* and *Biology and Philosophy*. He was awarded the Society for Marine Mammalogy's Fairfield Memorial Award for Innovation in Marine Mammal Research in 2001 and the National Center for Science Education's Friend of Darwin Award in 2003. Elsberry is involved with many online resources, including the Talk.Origins Archive (www.talkorigins.org), TalkDesign (www.talkdesign.org), and Antievolution.org (www.antievolution.org). He also hosts an E-mail list devoted to critiques of intelligent design.

ALAN D. GISHLICK is a postdoctoral scholar at the National Center for Science Education and has a Ph.D. in vertebrate paleontology from the Department of Geology and Geophysics at Yale University. Dr. Gishlick's research interests include functional transitions in the history of life, particularly the origin of avian flight within dinosaurs; the reconstruction of soft anatomy in fossils; systematics and taxonomy and their role in concepts of biological identity and uses in interpreting patterns of evolution; the teaching of the history of life; and the interface between science and religion, especially as it relates to biological evolution.

Gary S. Hurd received his doctorate in anthropology from the University of California, Irvine (1976). Initially involved in medical research, Hurd left the psychiatry faculty of the Medical College of Georgia to return full time to archaeology in 1986. His taphonomic research (research into the analysis of post mortem modification of bone) led to an involvement in forensic casework. A registered professional archaeologist, he has also been accredited as an expert witness in capital cases by the state of Tennessee and has consulted in data recovery and taphonomy for several law-enforcement agencies. He became involved in creationist arguments against evolution while working as the curator of anthropology and director of education for the Orange County Natural History Museum. A recipient of multiple academic honors, Hurd has recently returned to private industry as senior partner and principal investigator of Viejo California, an environmental consulting firm. He has an extensive publication record in medicine, archaeology, psychology, sociology, biology, and chemistry.

Istvan Karsai is a biologist and mathematician with East Tennessee State University, where he studies the emergence of complex patterns from the interactions of individual agents.

Gert Korthof is a biologist and software engineer with the National Institute of Public Health and Environment, the Netherlands, and since 1997 has operated the web site www.wasdarwinwrong.com, which is devoted to the study of creationist and noncreationist critiques of evolution and to alternatives to evolution.

Ian Musgrave is a molecular pharmacologist interested in the control of nerve activity. He has done postdoctoral work at the prestigious Institute for Pharmacology at the University of Berlin and was a Johnson & Johnson Young Investigator. After 10 years as a National Health and Medical Research Council researcher, Dr. Musgrave has been appointed as a senior lecturer at the University of Adelaide, where his work in molecular pharmacology regularly brings him into contact with evolution.

Mark Perakh, professor emeritus of physics, California State University, Fullerton, has written extensively on intelligent design. He is the author of four books and close to 300 scientific articles and is the recipient of many awards, including one from the Royal Society of London for the discovery and study of the photodeposition of semiconductor films. His latest book is *Unintelligent Design* (Amherst, N.Y.: Prometheus, 2004).

JEFFREY SHALLIT, professor of computer science, University of Waterloo, is editor-in-chief of the *Journal of Integer Sequences* and co-author of *Efficient Algorithms*, volume 1 of *Algorithmic Number Theory* (Cambridge, Mass: MIT Press, 1996) and *Automatic Sequences* (Cambridge, Eng.: Cambridge University Press, 2003).

NIALL SHANKS is a philosopher and a biologist at East Tennessee State University, where he studies the philosophical foundations of evolutionary theory. He is the author of *God, the Devil, and Darwin* (Oxford: Oxford University Press, 2003) and has debated with Michael Behe in the journal *Philosophy of Science*.

VICTOR STENGER is emeritus professor of physics and astronomy at the University of Hawaii at Manoa, adjunct professor of philosophy at the University of Colorado at Boulder, and president of the Colorado Citizens for Science. He is the author of five books; his latest is *Has Science Found God? The Latest Results in the Search for Purpose in the Universe* (Amherst, N.Y.: Prometheus, 2003).

DAVID USSERY is an associate professor at the Center for Biological Sequence Analysis, Technical University of Denmark. His review of Michael Behe's book *Darwin's Black Box*, posted on the web in 1997, has received one-half million hits.

Index